Developments in Primatology: Progress and Prospects

Series Editor

Louise Barrett

For further volumes:
http://www.springer.com/series/5852

Robert G. Bednarik

The Human Condition

Foreword by Dean Falk

Springer

Robert G. Bednarik
International Federation of Rock Art
Organisations
Caulfield South, VIC 3162
Australia
robertbednarik@hotmail.com

ISBN 978-1-4419-9352-6 e-ISBN 978-1-4419-9353-3
DOI 10.1007/978-1-4419-9353-3
Springer New York Dordrecht Heidelberg London

Library of Congress Control Number: 2011930515

Printed on acid-free paper

Springer is part of Springer Science+Business Media (www.springer.com)

This book is dedicated to my teachers, the gurda mirdanha, *men of the highest degree, in the Pilbara in northwestern Australia, who consented gracefully to introducing this benighted* walybala *to their wisdom.*

Foreword

Robert Bednarik is a renowned expert in archeology and prehistoric art whose ideas are synthetic, grounded solidly in science, and informed by a world perspective. He does not mince words when it comes to critiquing the field of Pleistocene archeology, or its diffusionist myth that modern humanity "developed in one privileged region of the world, and spread from there through a people incapable of interbreeding with any others." As I read *The Human Condition*, I occasionally found myself chuckling, and at other times gasping and thinking, "Did he really say *that*?" Who better than Bednarik to propose an iconoclastic hypothesis about human evolution as a replacement for the model that has dominated the field for decades? That's just what he has done in this book, which he predicts will be vigorously criticized by Anglo-American Pleistocene archeologists. I suspect that he is right. But, then, Bednarik did not write it for these specialists. Instead, he is targeting other kinds of scientists, and anyone who is fascinated with the question of how humans evolved to become what they are today.

With seven succinct chapters, *The Human Condition* is relatively short and engaging. It opens with a discussion of the history and philosophy of science that focuses on Pleistocene archeology. The author observes that this subfield has traditionally relied largely on tool types that are represented cross-culturally, rather than on more dynamic cultural customs that can shed light on the emergence of human consciousness. A persuasive case is made that archeological narratives about the early human past are probably largely false. One such narrative is the dogma that cultural change and evolution were relatively static before the advent of the Upper Paleolithic around 45,000 years ago, at which point there was a "quantum leap" or "explosion" in western and central Europe that resulted in the sudden emergence of the characteristics that distinguish modern humans—language, art, symbol systems, self-awareness, etc. A link is described between this short-range narrative and the "African Eve" model, according to which modern humans are all descended from one female *Homo sapiens* who migrated from Africa with a group of superior humans that, eventually, expanded and replaced all of the other hominin species. Thus, it was Eve's descendants who made all that nice Upper Paleolithic art. The author argues that "the Eve supporters have led the study of hominin origins on a monumental wild-goose chase," and that an accumulation of evidence supports the alternative multiregional model in which *Homo sapiens sapiens* evolved from more robust forms (e.g., Neanderthal-like) as a single breeding unit across Africa, Asia, and Australia.

Chapter 3 begins to amass "hard evidence" in support of a long-range gradualist hypothesis regarding the evolution of, not just human bodies, but also their minds. Forget the dogma about the sudden European Upper Paleolithic creative explosion. Much older evidence, most of which is not from Europe ("an insignificant appendage of Asia"), is reviewed in an illustrated discussion about the forms of symbolism that are embedded in paleoart. The reader is introduced to the world's oldest manuports, coloring pigments, rock art, petroglyphs, cupules, engraved portable objects, beads, and pendants. (In subsequent chapters, an interesting case is made that much of this art suggests a striving for perfection, which had implications for the evolution of associated neurological substrates. Bednarik suggests that paleoart is the main surviving indication of a proclivity for externally storing symbolic meaning—i.e., it formed a kind of prehistoric external hard drive.) We learn that "there is in fact far more surviving "Middle Paleolithic" rock art in the world then "Upper Paleolithic," which renders the claims that "modern" behavior patterns were introduced with the latter technological phase completely farcical." It is not surprising to read that the world's archeological community has a hard time accepting this evidence, and the accompanying discussion about academic power politics is gripping.

An analysis of early seafaring in Chapter 4 is a highlight of the book. Stone tools discovered on the Indonesian island of Flores show that the ancestors of *Homo floresiensis* had arrived there around 1 million years ago, which suggests that seafaring may have been developed in that part of the world, perhaps in conjunction with fishing. For a variety of reasons, the multiple crossings required to get to Flores from Bali were unlikely to have been accidental, and all would have entailed departures to opposite shores that were visible from the starting points. In order to shed light on the cognition and innovations required for seafaring, Bednarik participated in a First Mariners Project that constructed eight sea-going vessels to test the required technologies and to attempt sea crossings (four of which were successful). From these experiments, he surmises that maritime colonization of islands (and eventually continents) harnessed the forces of waves, wind, currents, and buoyancy. It required social cooperation, planning ahead, a concept of time, and probably technological inventions such as cordage and knotting. Bednarik suggests that seafaring would have promoted the formation of neurological changes supporting conscious awareness, symbolism, and a grasp of cause-and-effect relationships. He also makes a persuasive argument that it could not have happened without language.

Bednarik lays the groundwork for his own paradigm in Chapters 5 and 6 by focusing on significant physical changes that occurred in humans during the last 50,000 years. As people evolved into their modern forms everywhere, their skulls and skeletons became markedly less rugged and thick compared to those of earlier humans (e.g., Neanderthals), and their cranial capacities decreased. Some researchers (notably C. Loring Brace) associate this universal gracilization of humans with dental reduction attributed to cooking and other food-processing techniques. Although Bednarik accepts that cooking technology might have been a contributing factor more recently, he believes it could not have caused the general decrease in skeletal robusticity that occurred during the late Pleistocene. Instead, he suggests that the decrease in physical strength and brain size that accompanied modernization were deleterious, which leads him to conclude that the usual laws of biological evolution ceased to apply to humans during the last 50,000 years. Why would this be? Bednarik's answer is that human mating preferences became shaped by cultural ideals about sexual desirability (i.e., reproduction was acted on by sexual rather than natural selection), and that this changed breeding patterns. Specifically, men began to prefer women with childlike (neotenous) gracile features, which drove evolution much as the domestication and selective breeding of animals by humans altered their physiology and appearances. The gracilization of women was, thus, a trend that was incidentally channeled by male ideals of sexual attractiveness, while the trend for lessened robusticity in males lagged a bit, but eventually rode in on women's genetic coattails. Thus, Bednarik's maverick thesis is that "anatomically modern humans are the outcome of their own domestication."

Charles Darwin would have approved. In his 1871 book, *The Descent of Man, and Selection in Relation to Sex*, Darwin noted that a breed of animal (e.g., dogs) in which humans have selected for certain traits changes after several generations, and that two independent breeders working from the same basic stock will, over time, end up with animals that differ because each breeder will have impressed "the character of his own mind—his own taste and judgment—on his animals." Turning to humans, he observed great differences between men and women in features such as stature, muscularity, body shape, hair (beards, whiskers, moustaches), temperament, and voices, which he attributed to sexual selection (on both sexes) during human evolution. Unlike Bednarik, however, Darwin lacked access to a hominin fossil record (it hadn't yet been discovered), so focused on ethnographic variation rather than changes in hominin skeletons through time. He observed that, in humans, conventions of beauty varied markedly from culture to culture, which he attributed to the gradual accumulation in the past of different unconscious preferences that resulted in different outcomes for sexual selection in geographically-separated groups. Nonetheless, Darwin also speculated that more general secondary sex characters that appear in people, such as women being less hairy than men, evolved through sexual rather than natural selection at an extremely remote period. He also observed that the acquisition of such "a slightly injurious character" was not surprising because "we know that this is the case with the plumes of certain birds, and with the horns of certain stags." In other words, traits that would normally be eliminated by natural selection sometimes get a free ride on the coattails of sexual selection.

Bednarik goes much further in his final chapter, *Advanced human cognition: a Faustian deal*, and the result is an original and startling *tour de force*. Chapter 7 begins with a lucid review of the neuroanatomical features that distinguish human brains from those of other primates. Significantly, Bednarik points out that people are susceptible to a surprising number of neurological diseases not found in other primates (autism, schizophrenia, bipolar illness, Asperger's syndrome, etc.), and observes that these involve the most recently evolved parts of the brain. Further, "there is every possible indication that this neurological susceptibility is directly linked to...the ever-burgeoning brain...it has given us both the genius of our greatest thinkers and artists, and the despair of 'losing our mind'." This fits with Bednarik's self-domestication hypothesis, because domestication has long been known to promote unfavorable alleles. It also fits with his suggestion that the "futile search for perfection" entailed in paleoart, seafaring ("the pre-Historic equivalent of space travel"), the emergence of language, and the external storage of symbolic information was associated with a gradual change in human behavior from largely impulsive (as apes are) to obsessive. Thus, "obsessive-compulsive disorders seem to illustrate that obsessive and neurotic behavior is the price we pay for our rapid cognitive evolution"! Needless to say, Bednarik's gradualist hypothesis is very different, not to mention less flattering, than the short-range model that views human cognition as having emerged suddenly and relatively recently in one particular group of superior hominins. Bednarik predicts that "Pleistocene archeologists will, under the guise of testing them, move heavens and earth to disprove my propositions," and, no doubt, they will. But they have their work cut out for them because the book's essence is grounded in a voluminous amount of data regarding the latest findings in paleogenetics, paleoart, world-wide archeology, evidence for seafaring, paleoneurology, psychiatry, and much more. In sum, this book is a gutsy read about a fascinating subject that leaves no prisoners, and does not have a happy ending. It is provocative, important, and deserves to have "legs."

School for Advanced Research
Santa Fe, New Mexico

Dean Falk

Preface

He who despairs of the human condition is a coward, but he who has hope for it is a fool.
(Albert Camus)

The term "human condition" generally refers to what it means to be human and why we are the way we are. It is often expressed as the human predicament: our capacity for both good and evil, our "dark side," or the "troubled state and nature" of the human being. It can also refer to our limitations, such as our inability to go where only our imagination can take us, our futile yearning for everlasting life, or our never-ending endeavors to construct meanings where there are none. Encompassing all of the human experience, from the biologically determined events of our lives to the ways in which we react to or cope with these, the human condition can be perceived either as such a broad concept, or it can be more narrowly defined, be it in philosophical or in scientific terms. The concept was popularized by André Malraux's 1933 novel about the failed communist uprising of 1927 in Shanghai, *La condition humaine*, and his profound observation needs to preface the present volume: "In the realm of human destiny, the depth of man's questioning is more important than his answers."

All recorded human societies seem to subscribe to a universally accepted set of ideals of cooperation, love and altruism, which in some cases evolved in geographical isolation and is thus indicative of their universalism—and yet there can be no doubt that all these societies are also capable of great brutality, greed, hatred, and indifference to the suffering of others. This ambivalence accounts perhaps for the subconscious sense of guilt burdening us, nourished by religion. The human condition is defined by the feelings and emotions associated with our existence, our "conscious" experience of past and future, cognizance of the passage of time, and our vexed awareness of our mortality. It has led to a plethora of metaphysical questions and doubts, concerning the purpose of our existence, or that of the universe, to what happens to us upon death.

These are issues that can be dealt with, if perhaps reluctantly, by science, which at this stage in its evolution has no great difficulty in clarifying them satisfactorily. As a society we are, however, far from satisfied by these apparently nihilistic answers. Much the same will also apply to this book: after the last chapter, readers may experience a vague dissatisfaction with the answers it offers. These may not be what readers had hoped for; they may sound too unsophisticated, too artless in fact. Could the answers to our profound questions about our condition really be as simple as I contend on these pages? But is it not true that whenever there has been a major insoluble problem in the history of science, the answer, when finally found, has always been very simple?

In this book the human condition will be examined not as a metaphysical or ontological issue, but rather from a scientific perspective. However imperfect our comprehension of reality may be, the more we have learned to interpret the nature of ourselves in our configured reality construct, the less opinionated we have become—which is in a way encouraging. Where science differs so fundamentally from other constructs of the nature of the world is that it accepts not a single human claim of finite truth; all its interpretations are contingent formulations that may change in the future. This will be reflected in the first chapter, emphasizing the overwhelming immersion in misinformation that is also a part of the human condition, but is often overlooked. Born ignorant into a society that still has almost no understanding of the epistemology of its knowledge base, we are in no position to justifiably speculate about reality or talk of objectivity. But science has become very adept in teasing out innumerable minutiae about the nature of the world, and if these tiny snippets were correctly assembled across disciplines, they could collectively provide credible hints about how some of the major puzzles might be solved. The main disadvantage of modern science is the overspecialization it engenders, which may not be apparent at the low-level, mass-consumption academic environment, but which now creates almost unbridgeable gaps between disciplines at the high end, at the cutting edge. That inexpedience will be avoided in this volume, which will develop its theses from the perspectives of several disciplines, but especially from archeology, paleoanthropology, genetics, evolutionary theory, cognitive science, neuroscience, and clinical psychiatry.

As noted in the first sentence of this Preface *why we are the way we are* is crucial in gaining insight into the human condition. In this quest, no progress can be expected without contemplating the human past: what occurred in the development of our species and subspecies that determined an evolutionary trajectory resulting in such an unusual primate? It may not be sufficient to glibly note that it is our mindless superstition and pointless rituals that separate us from other animals; nor will simplistic references to "conscious thinking" or "self-awareness" suffice, because there is no satisfactory definition or explanation for either banality, and in the end they explain nothing. Bearing in mind that the framework required in considering our past is provided by two disciplines of questionable credentials, Pleistocene (Ice Age) archeology and paleoanthropology (the study of extinct forms of humans), provides an immediate obstacle to meaningful inquiry. Both these disciplines are subject to the vagaries of fashionable fads, erratic sectarianism, anthropocentrisms, academic fakery, and deference to authority, combined with an inherently poor susceptibility to falsification. Both present poor records of previous performance, and neither should be treated as science, in the proper sense of that word.

This state of affairs leaves us with the task of distilling from the accumulated "knowledge base" generated by these academic pursuits that part which can possibly be relied upon and which is relevant to our quest. This is obviously essential before we can consider how the human condition came to be what it is. In the following chapters it will be demonstrated that some of the most influential recent fallacies concerning human origins impact greatly on the research target chosen in this book, and that without teasing out these misconstructions, any notions relating to our primary subject, the origins of the human condition, would be illusory. In essence this involves interrogation of some of the most favored hypotheses of how and when human modernity evolved, and these are found not to meet any reasonable expectations we have of scientific propositions. Their empirical basis may evaporate upon close examination, their reasoning is often self-contradictory, and the ardency of individual protagonists in defending essentially unsupported theories can be disheartening. In this context the underlying problem soon becomes apparent: without some understanding of the emergence of what has been called "modern human behavior" it is impossible to know why we are the way we are, but the veracity of the information we have about this development is severely tainted by academic dogma. If this quest simply regurgitated what mainstream Pleistocene archeology offers us, the outcome would just be more mythology about our origins. Therefore, the first major task of this book is to clarify the status of the dominant hypotheses through critical and comprehensive review of the empirical data currently available.

This results not only in a significant reassessment; it invites dramatically different interpretations and syntheses, and a renewed endeavor to correlate ideas about the human past with the findings of innovative new approaches to what being human means. Once the framework of reference is refurbished in this way, it soon becomes apparent that the pertinent models of the relevant life sciences are much more readily reconciled with it. Indeed, as Malraux was quoted above, the significance is in the depth of the questions asked. This reassessment, impossible in the context of archeological and paleoanthropological dogma, prompts a suite of entirely new questions. One of the most consequential of them is: what could have caused the inherent laws of biological evolution to be suspended for humans during the last fifty millennia or so? And yet, this question has never been asked by the mainstream. Nor has the question of why evolutionary natural selection apparently failed to select against thousands of deleterious genetic predispositions and defects, ever been asked by those concerned with the human past. Yet neuroscientists have prominently and extensively asked and debated this issue. They have been unable to arrive at any credible answers, simply because the disciplines taking care of human evolution are lagging many decades behind those dealing with the workings of our brain, and are in fact largely still struggling in the dark. Nor has there been a sustained attempt to deal with such key issues as why it should be that the etiologies of brain illnesses suggest that they involve largely the same areas of the brain that are the phylogenetically most recent; or why it should be that other extant primates are largely if not wholly free of such pathologies. These are far more consequential issues to our origins than endless polemics about favored craniometric interpretations or stone tool knapping practices.

These inequities in the state of research are so profound that neuroscientist Todd M. Preuss has referred to *Homo sapiens* as "the undiscovered primate." A vivid demonstration of these inadequacies is provided by the greatest unsolved conundrum of evolutionary genetics: why evolutionary processes failed to select against the degenerative genetic predispositions of extant humans. The apparently quite recently developed toleration of maladaptive traits, which range from somatic features universally related to neotenization to mental disorder susceptibility alleles, and to almost countless other detrimental susceptibilities, remains supremely unexplored. Disorder susceptibility alleles have neither fixated, if adaptive, nor gone extinct, if maladaptive. Those that are rare, the single-gene Mendelian disorders, may escape selection precisely because they are rare, and the molecular bases of over 1700 of them have now been identified. However, the preservation of the mutations deriving from multiple mutant alleles at different genetic loci involved in the major deleterious etiologies has remained entirely unexplained until now.

Having been discussed at great length for a number of years, this puzzle has become the discipline's "unresolved paradox." The reason for this unsatisfactory state is very simply that the dominant hypotheses of recent human evolution, proclaiming replacement of all hominins by a new African species, render a solution impossible. If we account for *Homo sapiens sapiens* by either natural selection or genetic drift, as these hypotheses demand, there is simply no explanation for the neuroscientific paradox. A core purpose of the present book is to solve the paradox by replacing the old paradigm with a new one.

For the first time, a tenable and inherently plausible authentic solution is offered for the toleration of human neuropathologies by both natural selection and genetic drift. And for the first time, a credible explanation is presented for the extraordinary and sudden changes that led to "anatomically modern" people: the reduction in both brain size and somatic robusticity, as well as the loss of estrus, and many other features so crucial to appreciating what it is that made us what we are today. This book explains why the etiologies of brain illnesses suggest that the phylogenetically most recent areas of the brain are affected, which are the very same areas that underwrite our advanced cognitive abilities. It also explains the absence of neurodegenerative diseases in other primates, and why human males strongly prefer females presenting neotenous appearance and other features. And it offers clues for how, why and when dozens of mental illnesses and thousands of genetic impairments endemic to humans may have appeared. Without some appreciation of these issues such conditions have no causal context or explanation; we are merely trying to make sense of end effects at specific loci without an appreciation of how they came about. Science, however, expects some level of causal reasoning from us, and that is provided here.

In the realm of understanding the human animal, theoretical progress does indeed depend very much on the questions we ask, and there are many other questions asked in this book that were consistently eschewed by the mainstream. For instance, as the mental faculties of hominins increased with the rise to the top of the food chain, individual reliance was delegated to society and to objects, the latter ranging from tools to objects of storing symbolic information outside the brain ("exograms"). Orthodox Pleistocene archeology has shown little interest in the latter, designating them to categories comprehensible within simplistic reality frames of reference, for example, by defining them as "art objects." Moreover, it sustains a model linking the origins of these exograms to the advent of the purported African ancestors of our subspecies, another rather consequential bungle attributable to dogmatic intractability: the relevant empirical evidence to show the much earlier use of exograms has long been available, but was either ignored or explained away.

This book is obviously a wake-up call for some disciplines, while at the same time noting the patience shown by some others. The former have historically not taken kindly to such observations, having traditionally relied upon their inherently untestable status. Needless to say, this book is destined to be labeled "controversial" by the hegemonic disciplines it is critical of. However, the relevant hard sciences may take a different view and may be prompted into beginning to question the quality of the information and hypotheses provided by the humanistic gatekeepers of hominin history. If this should occur, a key purpose of this book would be achieved. It will not, however, bring about a paradigm shift in Pleistocene archeology; paradigm shifts do not occur in academic pursuits based on authority and internally unfalsifiable propositions.

Finally, those who believe in the exalted status of humans—that we share our image with a deity, or that as a species we have made this planet a better world—will categorically reject the key elements of this book. The biologically correct definition of our subspecies as a neotenous ape clashes significantly with their fantasies, which are simply another illustration of the human condition: we are prone to inventing grandiose self-delusions about our noble cause and character. Apparently, the truth about ourselves is so unbearable that we need to lie about it to ourselves, in the same way as we need to invent nirvanas and paradises, and beliefs of salvation. *Homo sapiens sapiens* truly is in need of some therapy.

Melbourne, Australia
August 2010

Robert G. Bednarik

Acknowledgments

Much of the original impetus for this book came from a series of eight lectures on *Cognition and symbolism in human evolution*, which I was invited to contribute to the Semiotics Institute Online, based in Toronto and run by Professor Paul Bouissac. I thank Paul for prompting me to gather my thoughts on this subject, from which eventually the concept of this book germinated. I also thank Professor Helmut Ziegert, Hamburg, for generously providing the image of the three Acheulian ostrich eggshell beads from El Greifa, Libya. Thanks are due to Professor Dean Falk, who graciously wrote the Foreword; and to Professor Ellen Dissanayake, for her valuable comments on the proofs. My gratitude is expressed to the 56 crew of my eight seafaring expeditions and to the several hundred members of construction teams involved in these replication experiments. In particular, I wish to single out the help and support of Peter Rogers, Emmanuel Littik (Om Ifé), Jacobus Zakawerus (Om Mberu), Bob Hobman, and Abdeslam El Kasmi. But most importantly, I welcome this opportunity to pay homage to my teachers who, in the 1960s, were then the oldest men of the Ngarluma, Yindjibarndi, and Njamal people of the Pilbara region of northwestern Australia; I thank them for teaching me to gauge the depths of my ignorance. I have kept my promises to them.

Melbourne, Australia — Robert G. Bednarik
October 2010

Contents

Chapter 1
A Little Epistemology

Prolegomenon

For a species described as intelligent, we have not been very successful in determining what it is that made us the way we are. The popular explanation of how we became human reflects a kind of consensus view forming the dogma that orthodox archaeology has created over recent decades. That dogma will be examined critically in this book, together with the epistemological currents that have formed its conceptual framework. One fundamental question may be a little perplexing to many readers: why do specialists consider human origins almost exclusively in terms of the somatic and technological development of our species? Surely "humanness" is not so much determined by the shape of supraorbital tori (brow ridges) or the mode of retouching flint tools. Surely what most distinguish us from other primates—and other animals—are a suite of distinctly "human" attributes: for instance, the ability to "store" symbolic information outside our brain; or our development of symboling capacities to such sophistication that they made it possible to modify our physical environment (or niche) on an extraordinary scale; and to harness its resources and energies in the ultimate quest of all species—the conversion of a significant part of the planet's biomass into themselves.

Bipedalism, for instance, is an adaptation widely found in the animal kingdom: all birds and many mammals manage to walk on two legs. Toolmaking is certainly not an exclusive preserve of hominins; many mammals and birds have this capability. The absence of estrus is also not unique to humans; it is shared by several other species. When it comes to communication, there is such a wide range of practices that most forms may still have to be discovered by science. From the amazingly complex messages of the octopus to the sign language capabilities especially of chimps, bonobos, and orangutans, virtually countless systems of communication are in use. The complexity of some of these may well match that of human languages and would probably exceed that of our nonverbal communication forms (e.g., body language). Indeed, the traditional separation between human and *other* animals has been gradually eroded throughout the twentieth century, most especially by the new science of ethology, to the point that separation now relies on just a very few criteria. And if we were to, for whatever reason, want to find out how we actually *became human*, our search would seem to benefit much from an initial focus on those rather

R.G. Bednarik, *The Human Condition*, Developments in Primatology: Progress and Prospects, DOI 10.1007/978-1-4419-9353-3_1, © Springer Science+Business Media, LLC 2011

few characteristics that appear to be uniquely and quintessentially human. It is a central purpose of this book to offer the means to attempt such a quest.

Today it is so profoundly self-evident that the biological sciences would be utterly impossible without Darwin's magnificent idea that we find it hard to imagine it otherwise. And yet, in appreciating Darwin's influence we may miss the most crucial aspect of it: that the great achievement of his idea was not that it presented the inevitability of evolutionary processes, but that Darwin was able to discover this despite the confining metaphysical system he had to exist in. After all, the knowledge that humans derive from other animals, that they are closely related to other species, has been with humanity for eons. The "ethnoscientists" of many pre-agrarian societies had long deduced this from their observations over tens of millennia. They had observed the similarities in skeletal details and arrangements of various organs and deduced the close affinities of all mammals. But for centuries, agricultural societies, rendered arrogant through their apparent ability to "master nature," had discounted such explanations as primitive myths.

One may well imagine the skepticism of the Australian Aborigines when missionaries focused their attention on them in the first half of the nineteenth century: their own, indigenous belief systems seemed to be so much more advanced and plausible. Then, in the second half of that same century, they learned that a great wise man in faraway England had discovered that humans descended from animals and are closely related to them. One could understand the Aborigines' frustration over the white man's misguided righteousness: what was it that prevented the Europeans from seeing the obvious, that they were conceptually backward? For these indigenes, and any other traditional societies, Darwin had merely reinvented the wheel, after religion had for millennia decreed that it needed to be square.

Apart from what are seen as cultural differences, people also differ in the way they construct relations between humans and nonhumans (cf. Descola's [2005] "analogism" versus "totemism"). Viveiros de Castro (1992) has explained that for the Amazonian Indians he has studied, human *culture* is what binds all living beings together, including other animals and plants, whereas they are divided by their different natures, i.e., their bodies. This view, in direct contradiction to the traditional European binarism of nature versus culture, appears the more advanced, not primitive. While Christians debated whether the "Indians" had souls, those same Amerindians applied principles of science when they experimented on conquistadors by drowning them, to see whether they would rot, thereby determining whether they had bodies. As Lévi-Strauss wryly observed, perhaps the Spaniards were better in the "social sciences," while the Amerindians conducted their research according to natural sciences protocol before Europeans had developed a science.

At this point, one might object along the lines that it might be a little overindulgent or careless to compare the observations of Holocene or Pleistocene ethnoscientists, the scientific observers of traditional societies, with the incredible sophistication that modern science brings to the task of comprehending the world. Before considering the epistemology of this latter science, on which we will reflect repeatedly, let us just consider, for the sake of illustration, a few relevant points.

For instance, our ancestors have been observing and collecting fossil casts for an incredibly long time span, as we shall see. We know from the "myths" of indigenous peoples how they sought to account for these and many other phenomena of nature, and they did so correctly occasionally. The many stories of great floods probably originate from such observations: humans have long sought to make sense of their surroundings, and the constructs of reality they created in the process still today determine how any group perceives the world—almost as if genetically imprinted in its members. Or consider another example: we know very well that our science's concepts of, for example, time and space remain most precarious. Some of those "traditional" explanations differ greatly from the consensus views Westernized societies have generated about these entities. But as we struggle with Heisenberg's uncertainty principle and cannot know whether Schroedinger's cat is dead or alive, we might benefit from being a little less certain that the ancients' answers must have always been less enlightened than ours. Our science is very sophisticated in a technological or "analogist" (*sensu* Descola) way, but when it takes on the trappings of a substitute religion it loses its edge instantly. In fact, it ceases to be science. And when we consider the possibility that our very reality is no more than *The imagined world made real* (Plotkin 2002) an abyss seems to open up: what are we to make of Western science?

Epistemology (from the Greek *episteme*, "knowledge," and *logos*, "theory"), or the *theory of knowledge*, is the branch of philosophy that deals with the nature and origins of knowledge. It addresses, among other things, standards or norms for justification and reasoning (including logic and probability theory), ideals of rationality, and the effects of specific philosophies (e.g., empiricism, relativism). As implied above, specific canons of rationality are thought to be historically contingent (Lewis 1929: 253; Mannheim 1929–1936: 57; Collingwood 1940: Ch. 6; Laudan 1977: 187) as well as culture-specific (Winch 1970: 97), and some Western authors have defined them as androcentric conflations biasing science in favor of male ways of experiencing the world. Descriptive epistemic relativism (e.g., deductive inference, causal reasoning; Swoyer 2002) has been improved in recent decades, but remains controversial. As historically and culturally situated creatures we cannot easily, if at all, step outside our concepts, standards and beliefs to appraise their fit with some mind-independent reality of Kantian "things-in-themselves." The trap of extreme relativism or postmodernism, already convincingly opposed by Plato (in his *Theatetus*), can also be avoided by *normative epistemic relativism*. It holds that while there are no framework-independent facts about the veracity of inference, justification, or rationality, there are facts about these variables relative to particular frameworks. Extreme relativism, on the other hand, invites solipsism: if one and the same thing can be true relative to one framework and false relative to another, true for some groups and false for others, there is no truth measure. This was countered by Plato thus: either the claim that truth is relative is true absolutely or else it is only true relative to some framework. If it is true absolutely, then at least one truth is not merely true relative to a framework, rendering the proposition apparently refuted.

A number of philosophers and social scientists (e.g., Quine 1960; Hollis 1967; Davidson 1984) have argued that we can only understand or interpret others if they largely agree with us about what is true, reasonable, justified, or the like. The academic endeavor has resulted in a variety of schools, the disciples of which are separated by "logical gaps": "They think differently, speak a different language, live in a different world" (Polanyi 1958: 151). Or to quote Kuhn:

> In a sense that I am unable to explicate further, the proponents of competing paradigms practice their trades in different worlds. . . . Practicing in different worlds, the two groups of scientists see different things when they look from the same direction (Kuhn 1970: 150).

Kuhn may have been unable to define what separates proponents of competing paradigms, but in the emerging "perspectivism" of Viveiros de Castro it may become possible to explain even the less mundane reasons (the mundane or obvious ones being individual intractability, jargon, professional isolation, *déformation professionnelle*, etc.). Indeed, these fundamental structural factors may herald one of the most severe challenges Western thought has yet been subjected to, in the way they challenge first principles and foundational assumptions. Just as the notion that the Western construct of reality must be valid because the West succeeded technologically is a falsity, the success of science is no proof that its present methodology provides ultimate explanations. Indeed, our science can be expected to look rather primitive in a thousand years, and our epistemology will appear as hopelessly naive as that of a millennium ago does today.

Be that as it may, some of the branches of the academic project have chosen to operate under a collective umbrella framework, called *science*; others have developed their own various frameworks. Science, today, favors a normative epistemic relativism over the notion of absolutism, but demands specific procedures of refutation and repeatability of experiments and strives for refutable theories cast in terms of causes. After all, quantum theory implies that determinism fails: objects need not always have determinate locations in space and time or determinate magnitudes (like a particular momentum or energy or spin). In all of this, the issue of testability of hypotheses is utterly paramount, involving two components: first, the logical property that is variously described as contingency, defeasibility, or falsifiability (which means that counterexamples to the hypothesis are logically possible); and second, the practical feasibility of observing a reproducible series of such counterexamples if they do exist. Thus, a hypothesis is testable if there is some real hope of deciding whether it is true or false of real experience. Relativism decrees that this does not render a discipline such as archaeology, which necessarily operates outside of falsifiability, in some way inferior; archaeology is simply an epistemic framework that has chosen to eschew scientific demands in favor of a different framework. Nevertheless, the bias in favor of science needs to be critically explored here before we embark on an investigation into the cognitive origins of our species.

What are the reasons for this preference for what appears to be a rather stilted scientific epistemology?

Anthropocentric Realities

Ignoring for the moment the differences between the realities perceived by different human societies, we can perceive two antithetical concepts of reality. One is of this wide range of realities as held by contemporary humans, with the perceptual and conceptual means available to them; the other relates the idea or abstraction of an "objective reality" (Kant's *"Das Ding an sich"*), which has been speculated to exist and which would have existed and would go on existing independent of human constructs of it. The former of these concepts cannot be a reflection of the latter (although it may reflect some aspects of it), and we need to appreciate that our examination of cognition or its epistemology is severely limited by the tools available to us: we can only study this phenomenon with means (conceptual constructs) that are its own, subjectively conceived products. This bootstrap approach may hardly seem a scientific basis (and it should be understood before we proceed), yet it is no less fragile than that of most other areas of "scientific" endeavor.

All phenomena of the physical, "real" world are made up of large numbers of variables, of which humans can only detect those which their sensory faculties, intellect, and scientific instruments allow them to perceive (Bednarik 1984: 29, 1985). From these they seek to select what I have called CCDs (crucial common denominators of phenomenon categories), which are the purported basis of all cosmological taxonomies. However, their selection is not determined by objective criteria in terms of how things really are in the world, but by the anthropocentrizing dynamics of human reality-building processes: by how phenomena can be interpreted and integrated into a system of understanding based entirely on human cognitive faculties. Since the latter were derived from human evolution, which was never in terms of defining cosmic reality, but in terms of such variables as survival value or procreational success, they must be assumed to provide at best a narrow spectrum of objective reality. Consequently, scientific constructs of reality should not be expected to adequately reflect real or objective reality (Bednarik 1985).

There are, however, exceptions. A phenomenon that is created by humans themselves, specifically for the purpose of relating to a human sensory faculty, can presumably only consist of those CCDs determining its phenomenological externalizations or reifications that are readily accessible to human perception. Art is such a phenomenon: there can be no CCD in art that is entirely inaccessible to humans. Indeed, art is the only phenomenon in the real world that provides human access to all of its crucial variables. One can invert this postulate by defining art (and a scientific definition of art is indeed badly needed) as the collective phenomena in human experience which we can behold objectively.

This truism explains how hominins attained their unique neural structures of relating to the world through paleoart—one of the core issues of this book. The introduction of phenomena consisting only of humanly perceptible variables, such as the production of symbolic surface markings, rendered perceived "reality" conceptually manageable, by providing complete rather than fragmentary sets of percepts. Visual and mental taxonomizing processes and the inclusion of the new neural structures in cybernetic feedback systems became thus available for evolutionary selection. "Conscious experience," or rather what we understand by it, became possible because the neural facilities prompted by certain activities, including earliest paleoart production, became available for the processing of stimuli of the nonartificial material world, in a taxonomizing format. This explains why the present results, humanly perceived "realities," were initially determined by factors such as paleoart and are in the final analysis both valid and inadequate. The cultural cosmologies or epistemological models they derive from are ultimately false, or at least significantly deficient, but there is no reason why a biologically intelligent species (Jerison 1973) could not form and maintain indefinitely such invented reality constructs (Bednarik 1990, 1994a).

" Biological intelligence" does not necessarily lead to a better grasp of objective reality for the species concerned. On the contrary, its development follows evolutionary laws and trajectories that render this unlikely, as they tend to lead an intelligent organism away from, rather than toward, objective reality. While it is true that intelligent forms of life must participate in a process that inevitably leads to the evolution of more intelligent forms of life, the improvements will always be in terms of their ability to enhance access to energy and nutrient resources and to promote procreational potential, never in terms of facilitating a better grasp of reality. Genotypes determine the sensory faculties of an organism, and changes only occur within the confines of phenotypic plasticity. These abilities determine which material stimuli an organism can detect. Genes can also form neural circuitry that allows cross-referencing of sensory information, but the ability to construct conceptual models of reality, which defines "intelligence" biologically, is not itself genetically determined. Among highly advanced life forms, selection will favor organisms capable of the conceptual and behavioral innovations from which new behavioral modes can be constructed: the mental faculties, *not the veracity of their constructs*, are the selective determinant.

Some years ago I proposed that "the evolution of our sensory facilities and intellect can be assumed to have only equipped us with adequate faculties to make them useful; they were not selected on the basis of their suitability in defining the reality of the cosmos—in fact *there was no survival value in that ability*" (Bednarik 1984). I argued that because of the limitations of the genetically based sensory systems of organisms, there could not be a direct correlation between humanly perceived reality and objective reality, and that this lack of relationship is the reason for the discrepancies between these realities. This, however, is not the only reason for the formulation of an anthropocentric world. There is at least one other, albeit more complex factor.

Human knowledge is derived from applying concept-building cognitive processes to external stimuli, i.e., sensory information, thus accumulating percepts. It is self-evident, I suggested, that human knowledge has a tendency to reinforce itself through its own products, because it is continually validated and augmented by our material and cultural achievements (Bednarik 1985). But this interdependence becomes rather more sophisticated and complete when we involve the role of culture. In the sense used here, the term "culture" does not refer specifically to human culture, but to the biological concept of culture: the individually acquired system of "understanding" which reflects the distinctive life trajectory of the organism in question (Handwerker 1989). In this sense, cultural dynamics refer to the processes by which the intelligent organism alters its perceptible environment through its dialectic participation in the processes shaping it. Selection in favor of increased levels of "intelligence" is the inevitable outcome of such interaction among percepts, concepts, and behavior patterns, but *at no stage of this autonomous process is there any need for the concepts to be in tune with objective reality*. Provided that the internally consistent logical framework is not challenged by it, there is no reason to assume that an entirely false, cultural cosmology or epistemological model could not be formed and maintained indefinitely by an intelligent species. Once again it is obvious that evolutionary success is irrelevant to the objective merits or validity of such models.

One can conjecture about the possible shortcomings of a cosmological construct, for instance, by comparing them to those of "scientific" constructs that are based on confirmation (Tangri 1989). Indeed, the comparison appears valid and quite illuminating. Just as the basic error in confirmation or induction is the inability of the inevitably subjective observer to identify the one variable of the phenomenon category that determines the common characteristics crucial to its operation (the CCD), the deficiencies of conceptual models of reality cannot possibly be explored from within such a model, which is the only way in which we have been able to proceed so far, as a species. In an anthropocentric system of reality, ideas or mental constructs must adhere to its inherent order not only to be acceptable, but even to be liable to be conceived—even though they comprise elements relating to material stimuli, i.e., elements that must be assumed to have some form of objective validity. This is because they can only be generated by involving memory traces based on the same system, and one could argue that the inherent order might simply be a reflection of neural hierarchies. Therefore, we cannot even determine whether there is any such order in objective reality "out there."

To summarize what has been said so far: the concepts of reality that have evolved in the course of hominin and human history have led to the range currently held by extant human populations. There is no reason to assume that these concepts could be particularly useful in exploring objective reality. What we can state categorically is that anthropocentrism governs all human consciousness, and that this knowledge has been with us for well over two millennia: it is quite clear from Plato's simile of the cave that he, for instance, subscribed to the concept of anthropocentrism. Developments in the twentieth century, in philosophy and theoretical physics, have begun to erode beliefs in the commonsense world epitomized by Euclid and Newton. The transient, ever-evolving nature of anthropocentrism has become more apparent since we have realized how many of even the supposedly most solid tenets have fallen by the wayside. As we approached the third millennium, we realized that the scientific enlightenment humans thought was within their grasp turned out to be a mirage, while the horizons of knowledge seem to be forever retreating from us. A Japanese saying summarizes it well: "As the radius of our knowledge increases, so does the circumference of our ignorance." Science itself continues to occupy a position that is for the most part within the human model of reality. It is, therefore, fundamentally subjective; many of its rules may be valid only within its own frame of reference. This is, of course, quite adequate in terms of the demands we make of it, as long as science is not expected to lead us to "objective reality."

These considerations render science's current preference for normative epistemic relativism very sensible indeed. If we lack access to objectivity, if our metaphysical construct provides only a contingent, tentative and interim model of reality, then it is most sensible to accept no givens and no finite certainties, but to explore the unknown world with null-hypotheses, falsifiable propositions, and a very skeptical mind. Let us remember that, if it were not for the color changes of ripening fruit, we may well lack color vision. As an intelligent creature with monochrome vision, we might one day discover that certain other species exhibit a strange ability of discriminating between identical shades of gray if these derive from differing wavelengths of reflected light. Not knowing that this is because they see color, we could not explain this phenomenon. We need to ask: how many phenomena, such as color, exist in reality, but are not detectable, by either our sensory abilities or their extensions, scientific instruments? We have not the faintest idea (remember Plato's prisoners in the cave!) but we can safely assume that our evolution has only equipped us with adequate faculties to render us "viable." This means, in effect, that our sensory equipment needs to broadly match that of the part of this planet's biomass which can serve as our food or whose food we might become. Just as bananas or cave lions have no need to comprehend reality, evolution has never equipped us for this role either. The human ascent, the subject of this book, is no doubt marked by human endeavors to comprehend the world, but the tools we brought to this task were hopelessly inadequate.

The humility implicit in this realization is where proper science begins.

The Role of Archaeology

Since these inherent inadequacies can safely be assumed to be related to the history of the early development of human consciousness, one potential course of illuminating them would be to focus on the period during which our cognitive niche might have been established. If valid information about the underlying processes could be secured, this might lead to the formulation of hypotheses about the origins of our anthropocentrism. While this would, of course, not by itself provide any access to objective reality, it might permit realistic inferences about the articulation between anthropocentric and objective reality. That some form of such an articulation does exist seems likely; it is difficult to see how there could be none at all. If one could explain the mental and cognitive processes involved, one might not only find it possible to consider the neural developments required and the biochemistry to account for them, but also find it possible, in the long term, to tackle the ultimate challenge: to explore reality outside that realm that is perceptually and conceptually accessible to humans. If one could explain how the cognitive basis of our world view was acquired—by quantifying, or at least defining, the processes involved—one ought to be able to speculate about the selective forces involved, how they contributed to the outcome, and how others would have altered the same.

The question of how much we know about the cognitive development of hominins, a key subject of this book, is obviously the starting point of any inquiry into these profound matters. How much do we actually know about the intellectual evolution of early humans, how reliable is it, what is it based on, and what are the reasons for the gaps in our knowledge? Clearly, this leads to questions concerning the epistemology of Pleistocene archaeology as we know it.

Archaeological studies, especially of the second half of the twentieth century, have been engaged in the *ecological deconstruction* of Pleistocene cultures. Yet the development of humanity is based on cultural and cognitive factors, not on genetically determined abilities to improve access to resources. Humans became human through processes enabling them to develop culture, cognition, and technology on a scale removing them far from all other primates in those areas. But Pleistocene archaeologists have preferred to view the ascent of our species as being recorded in natural processes that modified the skeletal structures of our ancestors. Indeed, humanness is widely seen as being expressed in cranial shape and other skeletal details, and archaeologists define the culture of early humans primarily through their perceived typologies of stone tools.

During the course of the late nineteenth and the entire twentieth century, paleoanthropology has made great efforts in illuminating the history of the physical (somatic) evolution of hominins. By comparison, relatively little effort has been directed toward learning about their cognitive and cultural evolution, and yet it would seem self-evident that it is not skeletal architecture or genetics that so much separates us from other primates, but the proliferation of our cultural and cognitive capacities. It is, therefore, quite right to say that the reasons for humanization and the processes involved have so far not been considered in adequate depth, and they have not been clarified. Indeed, the preoccupations of the disciplines of archaeology (which in the particular area of "cognitive archaeology" seems focused on a variety of relatively trivial issues, like shamanism) and paleoanthropology have led to research orientations that are so skewed that it would be unrealistic to expect them to be able to address the profound topic of cognitive evolution in a balanced fashion.

In the specific area of intellectual and cognitive evolution, ecological archaeology has provided us with only fragmentary, unreliable, and sometimes downright irrelevant evidence. Much of the discussion has centered on the human capacity to possess advanced language (itself an anthropocentric notion), and recent developments show to what a vast range of incompatible ideas an inappropriate research program can lead. To illustrate with an example: in respect of the so-called Neanderthals (and I use such a taxonomic pigeonhole only to conform with established terminology, without approving it; the reasons for my skepticism will become progressively apparent) we have the extreme views that on the one hand they were totally incapable of reflective language (e.g., Davidson and Noble 1989; Noble and Davidson 1996), and on the other they were capable of well-structured grammar and syntax (Falk 1987), and there are various intermediate views (e.g., Lieberman 1984). So in practical terms, Neanderthal's linguistic ability must lie somewhere between that of "an animal" and a "modern human"! One does not need archaeologists or anthropologists to arrive at such a view. Similarly, the beginning of "complex language" could be anywhere between 35,000 years and some millions of years ago, according to the various competing theories. This illustrates the impotence of Pleistocene archaeology in explaining much.

The types of evidence brought to the task of solving the problem include the cortical development inferred from cranial casts (Falk 1983), and the still continuing, unproductive speculation concerning the fossil laryngeal structure and the role of the hyoid bone (Marshall 1989), all of which is tenuous, far from unequivocal, and in view of the hyoid bone from Dikika (Alemseged et al. 2006) or the presence of the FOXP2 language gene in "Neanderthal" (Krause et al. 2007) probably irrelevant. But cortical or speech-related structures surely are results, not causes, of evolutionary selection favoring speech or intelligence: a selection criterion needs to be established before it can affect phenotypic selection of genes (Bednarik 1990). The reasons for the type of cortical developments we are interested in are not to be found in secondary symptoms, and by utilizing these in their hypotheses archaeologists have merely substituted symptoms for causes. What we need to ask is what could have been the true causes and dynamics in the cognitive developments that provided the new traits for selection, and which ones could have left detectable traces for us to discover? How would one find and identify such traces in the archaeological record?

As we will see in this book, there is ample archaeological evidence available to address these issues. However, instead of drawing from this rich source, Pleistocene archaeology has focused largely on creating a history of invented tool types. Tools, obviously, do not define cultures; they are cross-cultural artifacts, and in the sense the term is misused in archaeology it refers to Searle's (1995) "institutional facts." For instance, Searle explains, an object may be made partly of wood, partly of metal. Its property of being a screwdriver exists only because the person who makes or uses it represents it as such. Precisely, the same applies to an object made in the Pleistocene; it has factual properties and socially constructed, observer-relative properties. However, there is no evidence that the latter are shared between the ancient maker and user of the object, and its modern-day archaeologist interpreter. Nor do tools, or perceived assemblages of tools, define ethnic groups, social groups, language groups, tribes, nations, or civilizations. Moreover, the tools or other artifacts archaeology names are always etic or arbitrary constructs of "material evidence categories"; they do not define valid taxonomic entities, be it in the emic sense or in the sense of the CCD having been correctly identified. Therefore, even if diagnostic tools could define cultural traditions, it would still have to be doubtful that archaeology could distinguish these reliably. As we have observed above, a biologically intelligent group, such as Pleistocene archaeologists, can invent and maintain any cultural framework provided it is based on an internally consistent logic. In this case, it is not even testable, apparently.

The term "culture" defines several quite diverse entities. It can relate to (1) cells grown in an artificial medium containing nutrients; (2) the cultivation of plants or animals; (3) the customs, institutions, or achievements of a nation or other group of people; (4) the intellectual achievements (especially art) of humans collectively; or (5) a refined understanding or appreciation of culture in that last sense. In the sense the term is used generically in biology, it relates most closely to definition (4) and defines the passing on of practice by nongenetic means (i.e., by learning) and is, therefore, practiced not only by humans, but also by many other animals, especially primates (Handwerker 1989). Archaeologists sometimes use the term "cultural layer" to simply describe a sediment layer that contains charcoal, even if it contains no artifacts. There may be no proof that the charcoal in question is anthropogenic, in which case the term is fundamentally misleading. However, even in the presence of such artifacts as stone implements or pottery shards, the term "culture" is not appropriate. In the case of humans, "culture" defines the collective customs, beliefs, and arts of a group of people who are usually bound together by it, and these are passed on from generation to generation. It does not refer to tool types; we have no spear culture, knife culture, or can-opener culture. Tool types, obviously, exist cross-culturally, and to claim that certain specific archaeological "tool types" do define specific cultural entities is a circular argument. It is also an unfalsifiable proposition, hence not scientific, as indeed are all propositions of archaeology not based on data imported from the hard sciences. Similarly, the "cultural sequences" archaeology has provided for the Pleistocene exist only in the minds and writings of Pleistocene archaeologists; they may have no external or emic existence. Nobody would seriously propose, as an example, that there was ever a nation, tribe, people, ethnic group, or language group corresponding to the sequence of technological assemblages archaeologists define as the Aurignacian, and yet we speak of "Aurignacians" as if they were an ethnically, culturally, and historically definable reality.

That would be justifiable if they had a single universal culture, expressed, for instance, in rock art. But Paleolithic rock art specialists have demonstrated that they disagree vehemently about what constitutes Aurignacian cave art (see Clottes et al. 1995; Clottes 2001; Zuechner 1996; Pettitt and Bahn 2003; Valladas and Clottes 2003; Valladas et al. 2004, concerning the attribution of the Chauvet Cave rock art), and the ethnicity of these "Aurignacians" remains profoundly unknown (Bednarik 2007). So we define a "people" we think persisted for 15,000 years on the basis of certain combinations of tools, and then squabble over which rock art should be associated with these perceived tool traditions. Or, in other words, we use noncultural variables to define a culture and ignore cultural ones because we cannot agree which art "style" goes with which tool "style." That is a fair indication of the state of the discipline's epistemology: skewed research directions, inadequate methods, and misinterpretation of scientific data have all contributed to it.

In these circumstances, particularly when they are viewed from the perspective of taphonomic logic (Bednarik 1994b), it seems judicious to regard archaeological narratives of the earliest human past as probably being largely false. We will examine this proposition through the following example, which is particularly important to the core subjects of this book.

The "Mother of All Modern Humans"

In all fields, not only in archaeology, the dominant and the hegemonic can be both sustained and subverted by narratives (Ewick and Silbey 1995: 200). Narratives frame the world in a struggle for authority; they create ontologies. In the case of the Lower and Middle Paleolithic periods of human history, the dominant narratives of archaeology are more tenuous, more farfetched, and more invalid than for any other period of our existence as a genus. Over the past few decades, the dogma developed for these periods has become a caricature of archaeological interpretation. In its essence, this dogma perceives no cultural change or evolution throughout most of the Lower Paleolithic, roughly from 1.8 to 2.5 million years ago to 180,000 years ago. It defines this time as static and sees little change even in the subsequent "Middle Paleolithic" (better defined as Mode 3 tradition; Foley and Lahr 1997), which ends 40,000 BP in much of Eurasia, 20,000 BP in Africa, and only a few thousand years ago in Australia, or merely 200 years ago in Tasmania; Bednarik 2010). Then, with the advent of the Upper Paleolithic, around 45,000 years ago in some regions, the dogma perceives a cataclysmic "bottleneck," a "quantum jump," an "explosion": all the typically human characteristics that distinguish us from other animals appeared suddenly and at once—and, of course, in western and central Europe: art, language, complex social systems, self-awareness, forward planning, and symboling. This paradigm draws its inspiration largely from the "African Eve" model, according to which all living humans are the descendants of one single female. Her progeny lived in an unspecified part of sub-Saharan Africa in the late Middle Pleistocene, and for unknown reasons became genetically so different that they could no longer breed with other humans. Having asserted their intellectual and other superiorities over the neighboring peoples they began to expand, rapidly taking over the world as they eradicated or displaced all resident populations first in Africa, then in Asia, and finally in Europe. Upon reaching Southeast Asia around 60,000 years ago, they promptly started building seaworthy watercraft to continue on to Australia. By 30,000 years ago, they had colonized western Europe and were engaged in wiping out the resident Neanderthals completely, and they suddenly began painting in caves.

This is not, I emphasize, the absurd origins myth of some Californian religious cult. This is what most Anglo-American Pleistocene archaeologists believe actually happened, together with a good number of their colleagues elsewhere who agree with them. And this caricature is what is being taught in the universities of much of the world, but especially in Britain and the United States. This model has not one iota of archaeological evidence in its favor and was first invented by a German academic who has since been discredited as a charlatan. It is now based mainly on the speculations of some geneticists, opposed by other geneticists. Bearing in mind that the genetic divergence times based on unknown mutation rates, population sizes, and assumptions about colonization events are all dubious (Barinaga 1992; Templeton 1993, 1996; Ayala 1996; Brookfield 1997; Pennisi 1999; Strauss 1999), to say the least, it would appear that the formulation of the African Eve model was a simple misunderstanding. Perhaps the geneticists tailored their supposed mutation rates and other unknown variables to suit such emergence times for modern people they had been given to understand were reasonable, while the archaeologists assumed that the geneticists themselves had the correct numbers. Neither side effectively realized that the other was only guessing. In reality, population sizes as well as mutation rates and other crucial variables (such as the number of successful colonizations or influxes of genes in specific regions) are entirely unknown, and the divergence times given have no credible independent basis at all. Moreover, this "short range" or "replacement model" was based on numerous errors of fact and fake datings, as we will soon see.

The African Eve fad provides a classical example of an archaeological blunder that enthralled the discipline for decades and was absorbed by it to such an extent that it is currently still difficult to oppose it. It began with a series of fake datings of European human remains by Professor Reiner Protsch "von Zieten" (the aristocratic title is as bogus as the professor's results).

In the wake of public denouncement and a police inquiry (Schulz 2004), Protsch was sacked from his senior post at the University of Frankfurt. But two decades earlier, his fake results together with other unsound and since refuted information was used by German paleoanthropologist, G. Bräuer, to formulate his "Afro-European *sapiens* hypothesis," placing the origins of "anatomically modern humans" in Africa (Bräuer 1984). Indeed, his hypothesis was only a recycling of the African origins claim 11 years previously by Protsch himself (1973). In the late 1980s it in turn inspired the "replacement hypothesis," dubbed the "African Eve theory" by the media. British and American researchers formulated and keenly promoted the Eve model (Cann et al. 1987), after a team at the University of California at Berkeley had subjected 136 mitochondrial DNA samples to a computer program designed by Alan Templeton, attempting to construct a family tree for "modern humans." They reported that we must all descend from one common mother that lived about 200,000 years ago. Dr Templeton then pointed out that the same data could have generated 10^{267} alternative and equally credible haplotype trees (which is very much more than the number of elementary particles of the entire universe, about 10^{70}!), and the announcements were thus attributable to a computer bungle. Hartl and Clark (1997: 372) warn that no existing methods could guarantee that the most parsimonious tree result should even be expected to be the correct tree. Maddison (1991) had earlier demonstrated that a reanalysis of the Cann et al. (1987) model could produce 10,000 haplotype trees that were more parsimonious than the single one chosen by these authors. And whenever more parsimonious trees were found, their basal branches tended to include non-African types. Cann et al. had also misestimated the diversity per nucleotide (single locus on a string of DNA), incorrectly using the method developed by Ewens (1983) and thereby falsely claiming greater genetic diversity of Africans, compared to Asians and Europeans (they are in fact very similar: 0.0046 for both Africans and Asians, and 0.0044 for Europeans). Even the premise of genetic diversity is false, for instance, it is greater in African farming people than in African hunters-foragers (Watson et al. 1996), yet the latter are not assumed to be ancestral to the former (see, e.g., Ward et al. 1991). Many other objections have been voiced, among them the apparent morphological continuities in European and especially Asian hominin populations, and Alan Mann's earlier finding that tooth enamel cellular traits showed a close link between "Neanderthals" and present Europeans, which both differ from those of Africans (Weiss and Mann 1978).

Instead of abandoning their botched hypothesis, the Eve advocates tinkered with its details and continued promoting it so aggressively, through a few dominant journals, that it won rapid public approval, particularly in the English-speaking world. An important factor in the popularity of this model was the perception that it underpinned the idea of a single humanity, whose individuals are all ultimately related. But in fostering this feel-good notion of togetherness its academic backers overlooked two potential ideological objections: their tale of the rise of our ancestors who exterminated or out-competed all other humans on the planet involved a sinister side also (Bednarik and Kuckenburg 1999); and academic spin may foster academic careers, but it is detrimental to scientific veracity. At best, the claimed glorious triumph of our forebears would have come at a terrible cost to other humans; at worst it endorses fierce competition to the point of extinction and even becomes a rationalization of genocide. However, this ideology suited the sociopolitical climate of the 1990s ("Greed is good!"), and that is perhaps the main reason why such a highly unlikely paradigm became adopted as virtual dogma during that decade. Its opponents found themselves just as marginalized and reduced as the supposedly replaced robust humans. Indeed, only a handful of them had the fortitude to profess the obverse gradualist position through the 1990s.

The African Eve or replacement model is not the only version of an African origin of "anatomically modern humans." As noted, it was preceded by Bräuer's (1984: 158) "Afro-European *sapiens* hypothesis," and it was followed by, for instance, the "wave theory" (Eswaran 2002) and the "assimilation theory" (Smith et al. 2005). These variations differ significantly from the Eve notion, in that they concede the occurrence of genetic exchange between the purported robust and gracile populations. In other words, they acknowledge that "Neanderthals" and "Moderns" are conspecific because the definition of a species is that its members can produce fertile offspring. These alternative hypotheses are therefore only variations of the multiregional theory (Relethford 2001; Relethford and Jorde 1999), merely claiming an inflow of African genes. All models of a reticular gene flow are in fundamental agreement with Weidenreich's original trellis diagram (Wolpoff and Caspari 1996: 200–201; Bednarik and Kuckenburg 1999: 55–59); therefore they are ultimately of multiregional character (Fig. 1.1). Nevertheless, what unites all these models is that they assume a quantum jump in cognitive evolution in Europe at the time of the arrival of these "mythical Moderns" (Bednarik 2008) from Africa. The Eve model categorically excludes the possibility of any genetic contribution of robust humans once these "Moderns" had arisen.

Most Pleistocene archaeologists of the Anglo-American school also subscribe to the notion of a "big bang of consciousness," introduced by Richard Klein (Klein and Edgar 2002), that is supposed to have occurred at the beginning of a hypothetical period called the Upper Paleolithic. This is not a new idea; it was earlier expressed in John E. Pfeiffer's term "creative explosion" (Pfeiffer 1983). This explosion is variously thought to have occurred with the end of the Mousterian in Europe, with the disappearance of people called Neanderthals, with the beginning of a perceived technological phase called the Aurignacian, or with the appearance of people defined as anatomically modern. Even individual authors are most ambivalent about the timing of this "explosion of creative thought"; for instance, Steven Mithen places it variously at 30, more than 30 but less than 60, or at 50 millennia, in a single paper. He

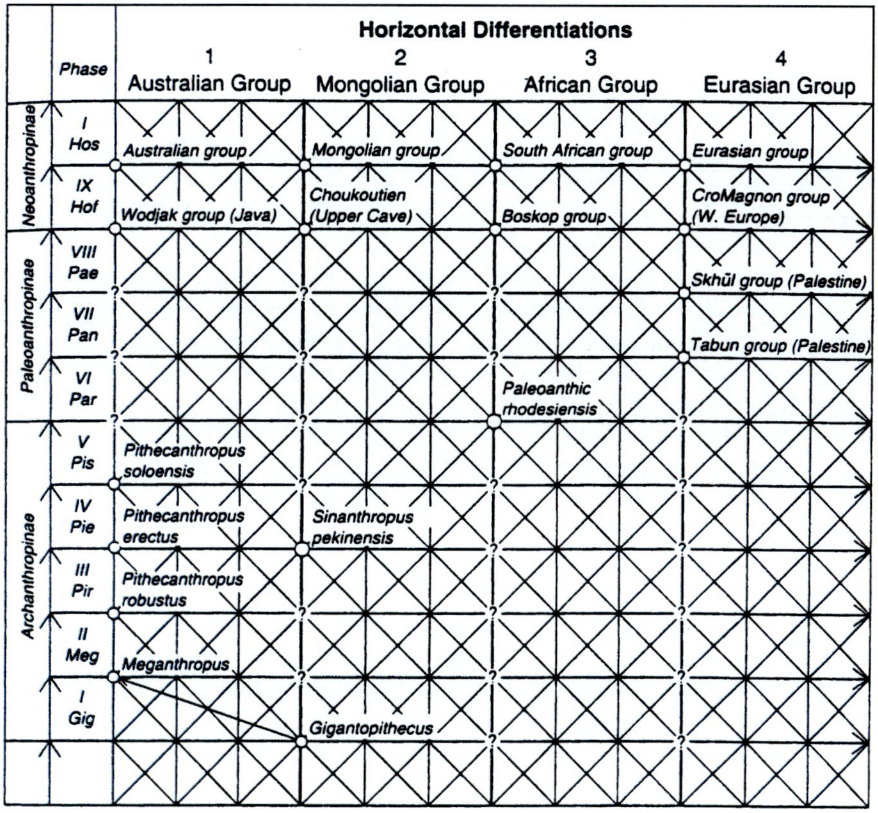

Fig. 1.1 Franz Weidenreich's "trellis" model of polycentric human evolution, of 1946

states quite explicitly (Mithen 1998: 168, 175) that it would be "perverse" not to describe the cultural developments from 60 to 30 ka (ka = thousand years) ago as a "creative explosion." Perversely he also decrees "to claim without doubt that art did not begin prior to 30,000 years ago is simply foolish" (Mithen 1998: 186). So he effectively presents us with two choices: to be labeled either foolish or perverse. He has since changed his tune so often one can hardly wait for the next capricious installment (his singing Neanderthals, for instance). It is clear that he has no idea what happened in recent human evolution, but—and I do apologize to him for singling him out as an example defining a widespread malaise in his field—this would be avoidable through greater familiarity with the relevant empirical evidence and better academic training: the English school is particularly backward in this field.

The problem with attributing this creative explosion to any specific time is that either these pronouncements are mutually contradictory or they refer to quite different points in time, which greatly detracts from their credibility. Yet despite their inability to pinpoint the timing of this decisive event in human history, the promoters of this idea stress that the "symbolic explosion model for the Middle–Upper

Paleolithic transition, criticized by Bednarik, has the merit of emphasizing the entirely modern character of the Aurignacian behavior" (d'Errico 1995: 618). The fantasy of this undated creative explosion can only be preserved if there are no finds of human creativity that predate its occurrence, which helps explain why its advocates are so coy about its timing. When confronted by individual finds that are substantially older than the "Upper Paleolithic," fundamentally challenging their model, they try to explain them away; or regard them as a "running ahead of time" (Vishnyatsky 1994); or pronounce them "untypical"; or challenge their dating or the scholarly competence of their presenters. This has been a consistent pattern for some decades—in fact similar accommodative reasoning in favor of whatever dogma rules at the time had been a persistent feature in all Pleistocene archaeology ever since all archaeologists loudly rejected the very idea of a Pleistocene archaeology in the mid-nineteenth century.

Archaeological Fetishes

However, since the end of the twentieth century, some of the advocates of these diffusionist ideas have begun to retreat from them. The number of supporters of the most extreme version, the replacement hypothesis, still espousing the Eve myth with any enthusiasm is now gradually dwindling. Based initially on fake datings of numerous human remains falsely attributed to the EUP (Early Upper Paleolithic) period, as it is called, this explanation is progressively being abandoned, for the many reasons we will visit in Chapter 2. But at its zenith the creative explosion scenario spawned some rather bizarre notions. For instance it led to the proposition that all humans prior to fully modern man should be placed with the apes rather than hominins (Davidson and Noble 1990). Countless quantified claims have been paraded concerning population sizes and densities, movements of people, and numerous other factors derived from the simplistic diffusionism this mode of thinking engenders. To illustrate this fad with one example: Webb (2006) presented an intricate model of how Asia was first colonized, involving 2820 people forming a single chain from Sinai to the Bay of Bengal, branching there into two beelines for Lantian in China and Modjokerto in Java. In all, Webb's pan-Asian relay race comprised 131 bands "ferrying fertile females to the spearheads" of his colonization parties. Although this deals with an event of colonization earlier than that of Eve's progeny, it illustrates the naive thought patterns of many archaeologists: they perceive, subconsciously perhaps, the maps of fossil finds as somehow informing us of population centers or population movements. It does not seem to occur to them that all archaeological finds of the Pleistocene are essentially attributable to preservation flukes, whose distribution is determined entirely by random processes, such as taphonomy and intensity of regional research efforts. This same pattern of sophisms, derived from an inadequate comprehension of what the "archaeological record" really constitutes, has been a significant factor in the origins myths expressed in the various hypotheses concerning the wandering tribes of "Moderns."

While the influence of these ideological currents appears to be waning at present, my concern is that, in accordance with traditional patterns, they will only be replaced with equally simplistic and half-baked models. As always in Pleistocene archaeology, there will not be a genuine paradigm change. Because of its enormous historical and hegemonic baggage, the discipline is incapable of radical reform. Pleistocene archaeology is a field that consists largely of blunders and mistakes, none of which have ever been fully resolved. From its very beginnings, before the middle of the nineteenth century, each and every major discovery or model has been resoundingly rejected by archaeology when first presented. This pattern, including the grudging and half-hearted acceptance after lengthy displays of indignation (especially as most such finds or improvements were offered by nonarchaeologists), continues to the present day. Pleistocene archaeology accepts corrections only after vigorously, and all too often viciously, opposing them, and only when the evidence becomes simply overwhelming. This practice is based on the falsity that the discipline already knows "what happened" in pre-History. If this were true we could close the book and abandon Pleistocene archaeology altogether; evidently it is not true.

This book seeks to address the question of how we as a species became "modern humans," a topic that is of considerable importance if we are to understand ourselves and our position in this world. It differs considerably from previous endeavors along these lines, including the most recent (e.g., Renfrew and Morley 2009). It is my considered opinion that Pleistocene archaeology, as it currently exists and operates, is only of limited use in such a quest. In attempting to demonstrate this point I have selected one of the most crucial periods in hominin evolution, that which witnessed the development of modern humans, however one defines them. The explanations the mainstream discipline offers for this are, I will propose, profoundly false. In these circumstances it would be impossible to develop any worthwhile theory about becoming human, on the basis of the current, and I fear future, dogmas of archaeology. If the subject of the origins of human modernity is to be explored in any productive fashion, it will have to contrast archaeological myths with the findings of the natural and cognitive sciences and other disciplines. Such a quest will have to be based on a complete and unbiased survey of all empirical evidence—a survey that is not guided by the preconceived notions of what would be *acceptable* findings. This will require us to free ourselves from any reified factoids of Pleistocene archaeology. In such a quest we need to first unlearn the dogma, the received knowledge of a discipline that has always been riddled with epistemological obscurities and conundrums.

Above we have briefly visited the concept of the CCD, the crucial common denominator of phenomenon categories. This is the common denominator of objective categories, in contrast to our imagined or anthropocentric denominators. If we consider critically all classes of taxonomic categories archaeology has invented for the Pleistocene, we find that they are inventions based on the etic perception of self-appointed experts. As we noted, there should not even be assumed to have been such a thing as a *people* that could be categorized as the Aurignacians; nor was there a *culture* corresponding to the Aurignacian. These are simply created concepts based on other "institutional facts" or observer-relative, social constructs of specialists who have no direct knowledge of these entities (Searle 1995). Nobody seriously believes (at least we hope not!) that one "Aurignacian" ever said to another something like "Pass me that Font Robert point, this waisted blade is unsuitable for my task." All tool types are merely archaeofacts or egofacts (Consens 2006); they have no independent existence outside the ideas and writings of Pleistocene archaeologists. Anyone who has discussed stone tools with traditional makers or users of them (as I have) knows that their emic nomenclatures of stone tools differ significantly from those created by archaeologists.

It follows that the same applies to all pigeonholes created by Pleistocene archaeology, because they derive ultimately from etic constructs that are likely to lack real validity. The entire sequence of Pleistocene "cultures" consists of nothing other than purported tool assemblages, comprising in most cases purported classes of lithics. Yet, as noted, tools do not define cultures. Consider, for instance, the term "Upper Paleolithic," which is so ingrained in our thinking about the human past that we can hardly communicate effectively without using it. Yet if we consider its defining separation from the subsequent "Mesolithic," we find that this discrimination must be at least partially an archaeofact. With the rapid rise of the sea level at the end of the Pleistocene period, the unknown half of Pleistocene humanity—that which had occupied the shores, lowlands, deltas, and lower courses of major rivers—became archaeologically visible for the very first time. Until that time, we know absolutely nothing about that half of humanity that was no doubt more sedentary and technologically more advanced than the mobile inland tribes following the herds. So the "advent" of the Mesolithic period simply marks the time when the coast dwellers enter the archaeological record. Therefore the definition of both periods is mistaken, unless the terms are used only in the broadest possible sense. (There are many other reasons why these definitions are false, too many to rehearse here.)

The only record we can have of Pleistocene shore dwellers would be from large inland lakes that have fallen victim to the lowering of aquifers in the late part of that period, such as the huge Pleistocene Lake Fezzan. Ziegert (2007a, b) reports that the Acheulian population on this former lake in the Libyan desert had established a village of stone huts (also reported elsewhere in the Sahara, e.g., in Morocco), buried their dead, and apparently used watercraft on the lake (Werry and Kazenwadel 1999). Yet orthodox archaeology claims sedentary settlements only begin with the Neolithic, hundreds of millennia later.

Similarly, the beginning of this "Upper Paleolithic" is also just an institutional fact, *sensu* Searle; it is not a fact, and all the "confirming" evidence in the support of the proposition of an abrupt change is no substitute for refutability. The divisions of the so-called Paleolithic period are completely arbitrary, having been established well over a century ago, and better ones would be possible in the light of today's knowledge (Bednarik 1995, 2002). Instead the discipline preserves and reinforces its outdated taxonomy. Much the same applies to designations of physical anthropology based on cranial and other fossil remains of individuals found. For instance, the term "Neanderthal" is used as if it defined a distinctive subspecies of humans, or even a separate species. But human remains that would fall within the morphological range regarded as Neanderthaloid occur in spatial and temporal contexts that exclude them from being Neanderthals. For instance, the Mesolithic skulls from Drigge and Hahnöfersand (Bräuer 1980; Terberger 1998; Terberger and Street 2003) are Neanderthaloid, as are those of many living humans, or of humans in Africa, eastern Asia, and Australia, i.e., outside the geographic range Neanderthals are supposed to have inhabited. Moreover, there are numerous known specimens that combine Neanderthal features with those of Graciles (Moderns) or, alternatively, earlier types. It is therefore perhaps more judicious to regard perceived Neanderthal features as valid variables, but not necessarily as the CCDs of a specific population—which is how they are usually seen.

What all of this is leading to is that we need to avoid treating taxonomic designations of the hominin history as reified facts, be they of human fossils, artifacts, or cultures. In reality, they are contingent formulations presented to be tested, to be falsified rather than confirmed by forcing every new find into a pigeonhole preconceived by premature projections of what should be valid categories. There simply are no valid categories in this entire field, of whatever entities it considers. What we do have are finds, such as human remains, that came to light in a totally random historical sequence over which the discipline exercised next to no control. There are enormous gaps in our incredibly patchy and sporadic record, and joining the widely separated dots it offers in time and space to predict some ultimate outcome is almost as premature today as it was a century ago. Any confidence in these practices, which are so central to the accepted routines of Pleistocene archaeology, is misplaced. This point is readily demonstrated by the pandemoniums that followed finds such as those from St Acheul, Neander Valley, Altamira, Trinil, Piltdown, Glozel, or, so very recently, Liang Bua on Flores.

Another factor to be considered is the relevance of a Eurocentric Pleistocene archaeology. Embedded in a contemporary European mindset is a predisposition of perceiving cultural evolution as Darwinian, progressing from the primitive to the developed. This clashes significantly with the observation that extant cultures in several continents would, under the European system, be defined respectively as Paleolithic, Mesolithic, Neolithic, Bronze, and Iron Age. In some countries, such as India, all of these "stages" coexist today with a Nuclear Age, which severely challenges the efficacy of this technological nomenclature in considering, for instance, the cultural or cognitive complexity of such societies. Clearly, a "Paleolithic" society can be as culturally complex as an "Iron Age" one, because the two sets of variables, technological and cognitive, exist largely independently. For instance, one might ask: are the Jarawas of the Andaman Islands (Sreenathan et al. 2008) who salvaged and reworked iron nails obtained from shipwrecks Stone Age or Iron Age people? Where is the Bronze Age of southern Africa, or how can the concept of a Neolithic be reconciled with the evidence from most of the world? Once the latter was intended to refer to the advents of pottery, ground stone artifacts, and plant and animal domestication, but we have recognized that none of these criteria are valid, except perhaps in a part of Europe. Decorated pottery in the Pleistocene Early Jomon of Japan and even earlier ground stone tools from Australia and New Guinea, together with the considerable complexity of the question of cultivation and domestication, render the traditional criteria of the Neolithic problematic, yet archaeology persists with this and many other superseded taxonomic pigeonholes and technological timetables instead of creating a universally valid, neutral nomenclature of *cultures*.

As noted above, this feature of archaeology is attributable to a fundamental misunderstanding about how cultures are expressed. The binarism of nature versus culture determining the operation of the neocolonialist discipline prompts it to use noncultural variables, especially stone tools, to define Pleistocene cultures, while ignoring cultural ones, such as paleoart, because we cannot agree which art "style" goes with which tool "style," when in fact both etic constructs are falsities. This already provides an inkling of the enormity of the task of effectively analyzing the discipline's epistemology. Similarly, we define the development of hominins primarily through their skull shapes—which is rather reminiscent of the practice of phrenology—instead of the characteristics that really determine humanness. This book will address this major imbalance, but only in the limited fashion possible in a single volume; the issues are too complex to consider comprehensively.

To tackle this labyrinthine task, it is particularly pertinent to first dismantle the mythology Pleistocene archaeology has created about human modernity having developed in one privileged region of the world, and having spread from there through a people incapable of interbreeding with any others. This diffusionism is a significant influence in the creation of archaeological mythologies. Early in the twentieth century, the fetishes—in the sense of objects representing something else, such as people—were, for instance, beakers. We invented the *Glockenbecher* people who wandered all over much of Europe, invading this and that region at whatever time. Today the fetishes are genes, representing populations that mass-migrated, for instance, from Africa to Europe. Let us examine how this contemporary fetish of Pleistocene archaeology fares when it is examined closely and critically.

References

Alemseged, Z., F. Spoor, W. H. Kimbel, R. Bobe, D. Geraads, D. Reed and J. G. Wynn 2006. A juvenile early hominin skeleton from Dikika, Ethiopia. *Nature* 443(7109): 296–301.

Ayala, F. J. 1996. Response to Templeton. *Science* 272: 1363–1364.

Barinaga, M. 1992. 'African Eve' backers beat a retreat. *Science* 255: 686–687.

Bednarik, R. G. 1984. The nature of psychograms. *The Artefact* 8: 27–32.

Bednarik, R. G. 1985. Editor's response to letter by B. J. Wright. *Rock Art Research* 2: 90–91.

Bednarik, R. G. 1990. On the cognitive development of hominids. *Man and Environment* 15(2): 1–7.

Bednarik, R. G. 1994a. On the scientific study of palaeoart. *Semiotica* 100: 141–168.

Bednarik, R. G. 1994b. A taphonomy of palaeoart. *Antiquity* 68: 68–74.

Bednarik, R. G. 1995. Traces of cultural continuity in Middle and Upper Palaeolithic material evidence. *Origini* 18: 47–67.

Bednarik, R. G. 2002. The human ascent: a critical review. *Anthropologie* 40(2): 101–105.

Bednarik, R. G. 2007. Antiquity and authorship of the Chauvet Cave rock art. *Rock Art Research* 24: 21–34.

Bednarik, R. G. 2008. The mythical moderns. *Journal of World Prehistory* 21: 85–102.

Bednarik, R. G. 2010. Australian rock art of the Pleistocene. *Rock Art Research* 27: 95–120.

Bednarik, R. G. and M. Kuckenburg 1999. *Nale Tasih: eine Floßfahrt in die Steinzeit*. Jan Thorbecke, Stuttgart.

Bräuer, G. 1980. Die morphologischen Affinitäten des jungpleistozänen Stirnbeins aus dem Elbmündungsgebiet bei Hahnöfersand. *Zeitschrift für Morphologie und Anthropologie* 71: 1–42.
Bräuer, G. 1984. The 'Afro-European sapiens hypothesis' and hominid evolution in East Africa during the late Middle and Upper Pleistocene. In P. Andrews and J. L. Franzen (eds), *The early evolution of man, with special emphasis on Southeast Asia and Africa*, pp. 145–165. Volume 69, Courier Forschungsinstitut Senckenberg.
Brookfield, J. F. Y. 1997. Importance of ancestral DNA ages. *Nature* 388: 134.
Cann, R. L., M. Stoneking and A. C. Wilson 1987. Mitochondrial DNA and human evolution. *Nature* 325: 31–36.
Clottes, J. (ed) 2001. *La Grotte Chauvet: l'art des origines*. Seuil, Paris.
Clottes J., J.-M. Chauvet, E. Brunel-Deschamps, C. Hillaire, J.-P. Daugas, M. Arnold, H. Cachier, J. Evin, P. Fortin, C. Oberlin, N. Tisnerat and H. Valladas 1995. Les peintures paléolithiques de la Grotte Chauvet-Pont d'Arc, à Vallon-Pont-d'Arc (Ardèche, France): datations directes et indirectes par la méthode du radiocarbone. *Comptes Rendus de l'Académie des Sciences de Paris* 320, Ser. II: 1133–1140.
Collingwood, R. G. 1940. *An essay on metaphysics*. Clarendon Press, Oxford.
Consens, M. 2006. Between artefacts and egofacts: the power of assigning names. *Rock Art Research* 23: 79–83.
Davidson, D. 1984. *Inquiries into truth and interpretation*. Clarendon Press, Oxford.
Davidson, I. and W. Noble 1989. The archaeology of perception: traces of depiction and language. *Current Anthropology* 30: 125–155.
Davidson, I. and W. Noble 1990. Tools, humans and evolution—the relevance of the Upper Palaeolithic. *Tools, language and intelligence: evolutionary implications*, pp. 1–21. Wenner-Gren Foundation, Cascais, Portugal.
d'Errico, F. 1995. Comment on R. G. Bednarik, 'Concept-mediated markings of the Lower Palaeolithic'. *Current Anthropology* 36: 618–620.
Descola, P. 2005. *Par delà nature et culture*. Gallimard, Paris.
Eswaran, V. 2002. A diffusion wave out of Africa. *Current Anthropology* 43(5): 749–774.
Ewens, W. J. 1983. The role of models in the analysis of molecular genetic data, with particular reference to restriction fragment data. In B. S. Weir (ed), *Statistical analysis of DNA sequence data*, pp. 45–73. Marcel Dekker, New York, NY.
Ewick, P. and S. Silbey 1995. Subversive stories and hegemonic tales: towards a sociology of narrative. *Law and Society Review* 29(2): 197–226.
Falk, D. 1983. Cerebral cortices of east African early hominids. *Science* 221: 1072–1074.
Falk, D. 1987. Brain lateralization in primates and its evolution in hominids. *Year-Book of Physical Anthropology* 30: 107–125.
Foley, R. and M. M. Lahr 1997. Mode 3 technologies and the evolution of modern humans. *Cambridge Archaeological Journal* 7: 3–36.
Handwerker, W. P. 1989. The origins and evolution of culture. *American Anthropologist* 91: 313–326.
Hartl, D. and A. Clark 1997. *Principles of population genetics*. Sinauer, Sunderland, MA.
Hollis, M. 1967. The limits of irrationality. *Archives Européenes de Sociologie* 7: 265–271.
Jerison, H. J. 1973. *Evolution of the brain and intelligence*. Academic, New York, NY.
Klein, R. G. and B. Edgar 2002. *The dawn of human culture: a bold new theory on what sparked the "big bang" of human consciousness*. Wiley, New York, NY.
Krause, J., C. Lalueza-Fox, L. Orlando, W. Enard, R. E. Green, H. A. Burbano, J.-J. Hublin, C. Hänni, J. Fortea, M. de la Rasilla, J. Bertranpetit, A. Rosas and S. Pääbo 2007. The derived FOXP2 variant of modern humans was shared with Neandertals. *Current Biology* 17(21): 1908–1912.
Kuhn, T. 1970. *The structure of scientific revolutions*, 2nd edn. University of Chicago Press, Chicago, IL.
Laudan, L. 1977. *Progress and its problems*. University of California Press, Berkeley, CA.
Lewis, C. I. 1929. *Mind and the world order*. Charles Scribners Sons, New York, NY.
Lieberman, P. 1984. *The biology and evolution of language*. Harvard University Press, Cambridge.

Maddison, D. R. 1991. African origin of human MtDNA re-examined. *Systematic Zoology* 40: 355.
Mannheim, K. 1929–1936. *Ideologie und utopie*. F. Choen, Bonn.
Marshall, J. C. 1989. Reply to P. Lieberman, J. T. Laitman, J. S. Reidenberg, K. Landahl and P. J. Gannon, 'Folk physiology and talking hyoids'. *Nature* 342: 486–487.
Mithen, S. (ed) 1998. *Creativity in human evolution and prehistory*. Routledge, New York, NY.
Noble, W. and I. Davidson 1996. *Human evolution, language and mind: a psychological and archaeological inquiry*. Cambridge University Press, Cambridge.
Pennisi, E. 1999. Genetic study shakes up out of Africa theory. *Science* 283: 1828.
Pettitt, P. and P. Bahn 2003. Current problems in dating Palaeolithic cave art: Candamo and Chauvet. *Antiquity* 77: 134–141.
Pfeiffer, J. E. 1983. *The creative explosion: an inquiry into the origins of art and religion*. Harper & Row, New York, NY.
Plotkin, H. 2002. *The imagined world made real: towards a natural science of culture*. Penguin Books, London.
Polanyi, M. 1958. *Personal knowledge*. Routledge, London.
Protsch, R. and A. Semmel 1978. Zur Chronologie des Kelsterbach-Hominiden. *Eiszeitalter und Gegenwart* 28: 200–210.
Protsch von Zieten, R. R. R. 1973. The dating of Upper-Pleistocene Subsaharan fossil hominids and their place in human evolution: with morphological and archaeological implications. PhD thesis, University of California, Los Angeles.
Quine, W. V. O. 1960. *Word and object*. MIT Press, Cambridge, MA.
Relethford, J. H. 2001. *Genetics and the search for modern human origins*. Wiley-Liss, New York, NY.
Relethford, J. H. and L. B. Jorde 1999. Genetic evidence for larger African population size during recent human evolution. *Journal of Physical Anthropology* 108(3): 251–260.
Renfrew, C. and I. Morley (eds) 2009. *Becoming human: innovation in prehistoric material and spiritual culture*. MacDonald Institute for Archaeological Research, Cambridge University Press, Cambridge.
Schulz, M. 2004. Die Regeln mache ich. *Der Spiegel* 34(18 August): 128–131.
Searle, J. R. 1995. *The construction of social reality*. Allen Lane, London.
Smith, F. H., I. Janković and I. Karavanić 2005. The assimilation model, modern human origins in Europe, and the extinction of Neandertals. *Quaternary International* 137: 7–19.
Sreenathan, M., V. R. Rao and R. G. Bednarik 2008. Paleolithic cognitive inheritance in aesthetic behavior of the Jarawas of the Andaman Islands. *Anthropos* 103: 367–392.
Strauss, E. 1999. Can mitochondrial clocks keep time? *Science* 283: 1435–1438.
Swoyer, C. 2002. Judgment and decision making: extrapolations and applications. In R. Gowda and J. Fox (eds), *Judgments, decisions, and public policy*, pp. 9–45. Cambridge University Press, Cambridge.
Tangri, D. 1989. Science, hypothesis testing and prehistoric pictures. *Rock Art Research* 6: 83–95.
Templeton, A. R. 1993. The 'Eve' hypothesis: a genetic critique and re-analysis. *American Anthropologist* 95: 51–72.
Templeton, A. R. 1996. Gene lineages and human evolution. *Science* 272: 1363.
Terberger, T. 1998. Endmesolithische Funde von Drigge, Lkr. Rügen — Kannibalen auf Rügen? *Jahrbuch für Bodendenkmalpflege Mecklenburg-Vorpommern* 46: 7–44.
Terberger, T. and M. Street 2003. Jungpaläolithische Menschenreste im westlichen Mitteleuropa und ihr Kontext. In J. M. Burdukiewicz, L. Fiedler, W.-D. Heinrich, A. Justus and E. Brühl (eds), *Erkenntnisjäger: Kultur und Umwelt des frühen Menschen*, pp. 579–591. Halle: Veröffentlichungen des Landesamtes für Archäologie Sachsen-Anhalt – Landesmuseum für Vorgeschichte, Vol. 57/2.
Valladas, H. and J. Clottes 2003. Style, Chauvet and radiocarbon. *Antiquity* 77: 142–145.
Valladas, H., J. Clottes and J.-M. Geneste 2004. Chauvet, la grotte ornée la mieux datée du monde. *À l'Échelle du Millier d'Années* 42: 82–87.
Vishnyatsky, L. B. 1994. "Running ahead of time" in the development of Palaeolithic industries. *Antiquity* 68: 134–140.

Viveiros de Castro, E. 1992. *From the enemy's point of view: humanity and divinity in an Amazonian society*. University of Chicago Press, Chicago, IL.

Ward, R. H., B. L. Frazier, K. Dew-Jager and S. Pääbo 1991. Extensive mitochondrial diversity within a single Amerindian tribe. *Proceedings of the National Academy of Sciences, USA* 88: 8720–8724.

Watson, E., K. Bauer, R. Aman, G. Weiss, A. von Haeseler and S. Pääbo 1996. MtDNA sequence diversity in Africa. *American Journal of Human Genetics* 59: 437–444.

Webb, S. 2006. *The first boat people*. Cambridge University Press, Cambridge.

Weiss, M. L. and A. E. Mann 1978. *Human biology and behavior: an anthropological perspective*. Little, Brown and Co., Boston, MA.

Werry, E. and B. Kazenwadel 1999. Garten Eden in der Sahara. *Bild der Wissenschaft* 4/1999: 18–23.

Winch, P. 1970. Understanding a primitive society. In B. R. Wilson (ed), *Rationality*. Harper & Row, New York, NY.

Wolpoff, M. and R. Caspari 1996. *Race and human evolution — a fatal attraction*. Simon & Schuster, New York, NY.

Ziegert, H. 2007a. A new dawn for humanity: lower Palaeolithic village life in Libya and Ethiopia. *Minerva* 18(4): 8–9.

Ziegert, H. 2007b. Report of the Hamburg Archaeological Mission to Budrinna (Fezzan, Libya). http://www1.uni-hamburg.de/helmut-ziegert/pdf/projects/permanentsettlement/HZiegert-BudrinnaReport(2007).pdf

Zuechner, C. 1996. The Chauvet Cave: radiocarbon versus archaeology. *International Newsletter of Rock Art* 13: 25–27.

Chapter 2
The Expulsion of Eve

It has been said that though God cannot alter the past, historians can; it is perhaps because they can be useful to Him in this respect that He tolerates their existence
(Samuel Butler)

Reviewing the Corpus in Question

This is the incredible story of how a couple of decades ago an entire academic discipline fell under the spell of an inherently improbable hypothesis, the advocates of which succeeded in neutralizing almost all academic dissent; and how it eventually, against considerable odds, began to unravel. The conditions required to sustain the African Eve hypothesis are quite obvious and precise:

1. Because Graciles and Robusts are purported to be different species, they need to be sharply separated in the paleoanthropological record by distinguishing morphological features.
2. For the same reason the genetic signatures immediately before and after the claimed replacement need to be sufficiently different to indicate separate species.
3. As the claimed victory of the Graciles over the Robusts is attributed to the formers' superior technological capabilities, there needs to be clear evidence of different technologies before and after the "replacement."
4. Another factor supposedly separating the Robusts from the Graciles replacing them is the "big bang" introduction of paleoart, i.e., apparently symbolic productions reminiscent of "art." Therefore, such evidence should only occur with the Graciles.
5. Since the Graciles are claimed to have come from sub-Saharan Africa, and traveled via the Levant and southeastern Europe, it would be expected that evidence of their presence can be found first in their homeland and later progressively along such a route, in the form of the arrival of a dramatically different technology as well as paleoart.

R.G. Bednarik, *The Human Condition*, Developments in Primatology: Progress and Prospects, DOI 10.1007/978-1-4419-9353-3_2,

In testing these reasonable and crucial propositions against the evidence I begin with the first. When I noted that we lack any skeletal evidence of the people of the Early Aurignacian (Bednarik 1995a), African Eve supporter Randy White chastised me:

> [T]he idea that there is no skeletal evidence to suggest that the Aurignacian was the work of anatomically humans is overdrawn. . . . Bednarik seems to have forgotten the modern human crania from Aurignacian sites like Vogelherd, Cro-Magnon, and Mladeč (White 1995).

An examination of this reaction would serve as a heuristic device to determine why such fads as the Eve model (and many others) can gain currency in Pleistocene archeology so easily. The three sets of human fossils White cites are among those most often listed by the replacement advocates as demonstrating the full anatomical modernity of the "Aurignacians." However, all three sets are in fact not of an age to place them in this technological pigeonhole, and at least in the case of the four Stetten specimens from Vogelherd (in the Swabian Jura, southwestern Germany) it had always been perfectly transparent that they were much younger, deriving from intrusive Neolithic interments (Czarnetzki 1983: 231; Gieseler 1974). That attribution has since been squarely confirmed by Conard et al. (2004): direct carbon isotope determinations, of samples taken from the mandible of Stetten 1, the cranium of Stetten 2 (Fig. 2.1), a humerus of Stetten 3 and a vertebra of Stetten 4, all agree, falling between 3980 ± 35 BP and 4995 ± 35 BP. Why all the Eve advocates (e.g., Protsch 1975; Bräuer 1981, 1984a, b, 1989: 136; Stringer 1984a, b, 1985, 1989; Stringer and Andrews 1988; Mellars and Stringer 1989; Wainscoat et al. 1986; Wainscoat 1987; Cann et al. 1987), and even others, such as Churchill and Smith (2000a, b), were unaware of these serious doubts is a mystery.

The Cro-Magnon sample White cites, derived from four adults and three or four juveniles, had been subjected to so much pseudoscientific spin that separating it

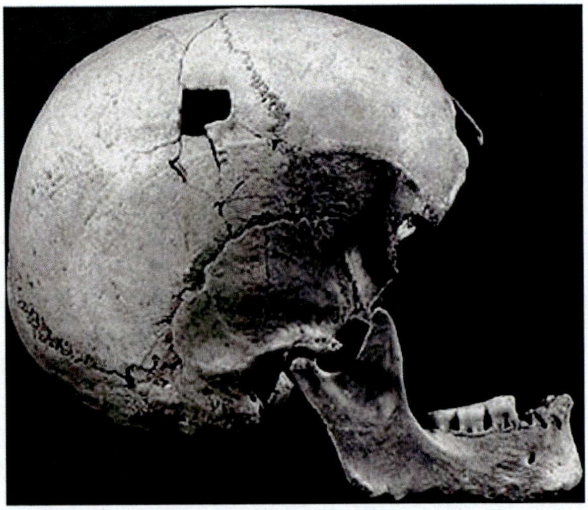

Fig. 2.1 The Stetten 2 skull from Vogelherd, Germany, attributed to the Aurignacian; the Stetten human remains are in fact all of the late Neolithic

from credible accounts is not readily possible. It had somehow become the type fossil of all "anatomically modern humans" (see Tobias 1995 for a most pertinent critique of this vacuous and blatantly anthropocentric term), making all contemporary people "Cro-Magnons." The group is in reality quite robust, and especially the very pronounced supraorbital torus, projecting occipital bone and other features of cranium 3 are Neanderthaloid rather than gracile. Sonneville-Bordes (1959) placed the sample from the Cro-Magnon shelter, just outside Les Eyzies, in the late Aurignacian; Movius (1969) suggested an age of about 30 ka BP and preferred an attribution to the Aurignacian 2. Both opinions, and numerous others, including White's, are refuted by the direct dating to about 27,760 carbon years BP (Henry-Gambier 2002); it places the Cro-Magnon individuals in the Gravettian rather than the Aurignacian technocomplex.

White's third example, that of the Mladeč specimens, is even more tenuous. There is no clear evidence that Pleistocene humans ever entered this cave, partly excavated about 130 years ago. Most of the macro-faunal remains in it apparently fell through the large shaft in the cave's roof, and Smyčka (1922: 118–119) proposed that the human remains had also dropped through this chimney, which is probably the case. The first group of documented archeological materials originates from J. Szombathy's second digging season, in 1882. The center of the Dóm mrtvých (Dome of the Dead) yielded in the upper part of the sediments twenty-two perforated animal teeth (probably of a single necklace), a long bone point, several fragments of points or awls, a utilized lower jaw of *Ursus spelaeus*, and two flint artifacts (Szombathy 1925: 8). In the subsequent decades the cave became a quarry for phosphate loam, and Knies (1906) reports that there were scattered and trampled bones along the road leading to the top of the Třesín Hill. In 1904 a small quarry was opened 20 m west of the entrance to the main cave and the sediments of the small horizontal passage were quarried (Knies 1906; Smyčka 1907). It is impossible to determine the find spots of the five bone points from Jan Knies' collection as there is no mention of them in his records (Szombathy 1925: 9). Little is known of the clearing operations of the Museum Society in Litovel from 1911 to 1922, or of the subsequent excavations by J. Fürst, E. Smékal, H. Rohm, and others (Fürst 1922, 1923–1924; Smyčka 1922, 1925). It is impossible to locate the sites where the human remains were found, except for a mark made by Rohm on a cave plan (in Weiser 1928: 281, Point 3). Most archeological finds have been lost (e.g., lithics and all bones published by Szombathy 1925: figure 9; and the fragments of "bone awls" from the old exposition near the cave, cf. Skutil 1938: Note 76). Skutil (1938: 32, figure 60) found two "Aurignacian" stone tools in the loess of the cave entrance and mentioned that J. Novotný discovered a blade core below the chimney (1938: Note 77). The excavations by the Moravian Museum from 1958 to 1961 located no evidence of any Upper Paleolithic occupation within the cave, and the view developed that the cave entrance had become sealed prior to the Würm Glacial (Jelínek 1987; Svoboda et al. 2002). It was guided by a comparison with Koněpruské jeskyně (Zlatý Kůň) in the Bohemian karst, where similar circumstances pertain (which, however, is now thought to be "Magdalenian"). The paucity of stone implements in the interior of Mladeč Cave, compared with the vast quantities of faunal remains, is conspicuous, and it needs to be emphasized that there is no sound evidence linking the apparently Aurignacian artifacts with any of the human remains. Nor does the assumption that "Aurignacians" occupied the cave derive any support from the hypothesis that the red pigment marks in its interior (Oliva 1989) are Paleolithic rock art. Of the sixteen red marks, four are definitely of the nineteenth century and the rest almost certainly so too (Bednarik 2006). Finally, the dating obtained from the reprecipitated calcite on the wall 7 m west of point "a" (of about 34 ka BP) is much older than the dates later obtained directly from bones and has been questioned (Svoboda et al. 2002).

In view of the lack of credible stratigraphic evidence from the site, the recent attempt to provide direct dates from some of the human remains is of greater relevance (Wild et al. 2005). A series of dates derived from specimens Mladeč 1, 2, 8, 9a, and 25c range from about 26,330 BP (the ulna of 25c) to 31,500 BP. It is, therefore, at best, partly of the very final phase of the Aurignacian period with its duration of about 15,000 years. More likely, most or all of the series is of the Gravettian technocomplex. Moreover, there is considerable evidence that the Mladeč specimens were far from "fully modern" (Smith 1982, 1985; Frayer 1986; Trinkaus and Le May 1982; and especially Jelínek et al. 2005). Notably, there appears to be pronounced sexual dimorphism, with male crania characterized by thick projecting supraorbital tori, Neanderthaloid posterior flattening, low brain cases, and very thick cranial vaults—all typical robust features. As in "Neanderthals," cranial capacities exceed those of Graciles (1650 ml for Mladeč 5), but there is a reduction in the difference between male and female brain size relative to Neanderthal data. The dimorphism is also expressed in the more inclined forehead in the males, their more angled occipital areas with lambdoidal flattening, broad superior nuchal planes, and more prominent inion. The female specimens show similarities with, as well as differences from, accepted Neanderthal females, such as larger cranial vaults, greater prognathism, lack of maxillary notch, a very narrow nose, and distinct canine fossa. However, the females are more gracile than the males, while still being more robust than males of later periods. The Mladeč population thus seems to occupy an intermediate position between late Neanderthaloid *Homo sapiens*, and *H. sapiens sapiens*, a position it shares with numerous human remains from other Czech sites.

This is an important issue we will return to later in this book. Suffice it to note here that the material from Pavlov Hill, an important Czech site, is among the most robust available from the European Upper Paleolithic, sharing its approximate age of between 26 and 27 ka with yet another Moravian site of the Gravettian, Předmostí. The more gracile finds from Dolní Vestonice are around 25 ka old and still feature some archaic characteristics (particularly the Neanderthaloid specimen DV16). Another find that has been considered as very early European "Modern" is the calotte from Podbaba, near Prague, variously described as sapienoid and Neanderthaloid, but undated; it probably belongs to the Mladeč-Předmostí-Pavlov-Dolní Vestonice spectrum. Morphologically similar specimens also come from Cioclovina (Romania), Bacho Kiro levels 6/7 (Bulgaria), and Miesslingtal (Austria), so this is unlikely to be a local phenomenon. Indeed, it needs to be seen in the greater Eurasian context.

Randy White's hasty comment can also be placed in a greater context, by reviewing the generic issue of the ethnicity of other European specimens of the period linking the decline of Mode 3 technologies (called "Middle Paleolithic," cf. Foley and Lahr 1997) with the final phase of the Late Pleistocene (called "Late Upper Paleolithic"). Turning next to adjacent western-central Europe, the extraordinary recent developments in German paleoanthropology are of considerable relevance here. Not only is there the correction to the age of the "robust" Neolithic human remains from Vogelherd, which the mitochondrial Eve advocates had been all too keen to place at 32 ka; nearly all of the German fossils claimed to be of the Upper Paleolithic are now thought to be of the Holocene. Of particular interest is the Hahnöfersand calvarium, described as so robust that it was judged to show typical Neanderthal features (Bräuer 1980) and hailed as the northernmost Neanderthal found. It was initially dated to the earliest "Upper Paleolithic" (Fra-24: 36,300 ± 600 BP; UCLA-2363: 35,000 ± 2000 BP, or 33,200 ± 2990 BP; Bräuer 1980), which conflicts sharply with results secured by Terberger and Street (2003): P-11493: 7470 ± 100 BP; OxA-10306: 7500 ± 55 BP. The re-dating of the skull fragment from Paderborn-Sande yielded even more dramatic differences. Originally dated at 27,400 ± 600 BP (Fra-15; Henke and Protsch 1978), Terberger and Street (2003) report an age of only 238 ± 39 BP (OxA-9879). In fact the skull was so fresh that it emitted a putrid smell when Terberger and Street drilled it for sampling. Then there is the cranial fragment of Binshof near Speyer, dated by R. Protsch in the 1970s as Fra-40 to 21,300 ± 320 BP. According to Terberger and Street it is only 3090 ± 45 carbon years old (OxA-9880). These authors also analyzed two individuals from the Urdhöhle near Döbritz, which had been attributed to the Upper Paleolithic, and found them both to be about 8400 years old. Indeed, of all the German "Upper Paleolithic" human remains, only one remains safely dated to earlier than 13,000 BP: the interred specimen from Mittlere Klause in Bavaria. A carbon isotope date of 18,200 ± 200 BP (UCLA-1869) from a tibia fragment (Protsch and Glowatzki 1974) has been confirmed by Terberger and Street's date from a vertebra, of 18,590 ± 260 BP (OxA-9856). It has, therefore, become clear that there are currently no "modern" remains from the first half, if not the first two-thirds of the west-central European "Upper Paleolithic." Nearly all the dates for German humans from the radiocarbon laboratory of the University of Frankfurt am Main appear to be substantially false, as do some of those from the University of California, Los Angeles. In addition, another German key specimen, the skull from Kelsterbach, has mysteriously disappeared from the safe of the Frankfurt institution. It had been dated to 31,200 ± 1600 BP (Fra-5) (Protsch und Semmel 1978; Henke und Rothe 1994), but is also believed to be of the Holocene, perhaps the Metal Ages (Terberger and Street 2003).

Then there are the robust but "modern" hominin remains of the EUP ("early Upper Paleolithic") at Velika Pećina, Croatia, close to the Neanderthal site Vindija. This specimen, too, has been a principal support for the replacement advocates, but it has also joined the long list of European humans whose age was grossly overestimated. It is now considered to be only 5045 ± 40 radiocarbon years old (OxA-8294; Smith et al. 1999).

The currently earliest, liminal "intermediate" finds in Europe, the Peştera cu Oase mandible and face from southwestern Romania (Trinkaus et al. 2003; Rougier et al. 2007), are perhaps about 35,000 radiocarbon years old, but they are without an archeological context. Although in some aspects "modern," the "derived Neanderthal features" of the mandible include cross-sectional symphyseal orientation, exceptionally wide ramus, exceptionally large third molars, and unilateral mandibular foramen lingular bridging. The partially preserved facial remains found in a different part of the extensive cave system and apparently from another individual, also combine robust and gracile features. More recently, Soficaru et al. (2006) have reported six human bones from another Romanian cave, Peştera Muierii, also clearly intermediate between robust and gracile Europeans. Found in 1952, they have now been dated to about 30,000 carbon years, which might correspond to around 35,000 sidereal years, and combine a partly modern, partly archaic brain case with a suite of other intermediate features.

The loss of the only relevant Spanish remains, from El Castillo and apparently of the very early Aurignacian technocomplex, renders it impossible to determine their anatomy. French contenders for EUP age present a mosaic of unreliable provenience or uncertain age, and direct dating is mostly not available. Like the Vogelherd and other specimens, those from Roche-Courbon (Geay 1957) and Combe-Capelle (originally attributed to the Châtelperronian levels; Klaatsch and Hauser 1910) are now thought to be of Holocene burials (Perpère 1971; Asmus 1964), and the former is now apparently lost. Similar considerations apply to the partial skeleton from Les Cottés, whose stratigraphical position could not be ascertained (Perpère 1973). Finds from La Quina, La Chaise de Vouthon, and Les Roches are too fragmentary to provide diagnostic details. The *os frontale* and fragmentary right maxilla with four teeth from La Crouzade, the mandible fragment from Isturitz, and the two juvenile mandibles from Les Rois, about 28–30 ka old (Ramirez Rozzi et al. 2009), range from robust to intermediate (e.g., Trinkaus 2007). Just as the Cro-Magnon human remains now appear to be of the Gravettian rather than the Aurignacian, so do those from La Rochette. The Fontéchevade parietal bone does lack prominent tori (as do many other intermediate specimens) but the site's juvenile mandibular fragment is robust.

This pattern of features intermediate between what paleoanthropologists regard as Neanderthals and Moderns is found in literally hundreds of specimens apparently in the order of 45–25 ka old (including the large Czech collection lost in the Mikulov Castle fire at the end of World War II). They occur in much of Europe, and intermediate forms between archaic *Homo sapiens* and *Homo sapiens sapiens* existed also in Asia and Australia. They include examples from right across the breadth of Eurasia, such as those from Largo Velho, Crete, Starosel'e, Rozhok, Akhshtyr', Romankovo, Samara, Sungir', Podkumok, Khvalynsk, Skhodnya, as well as Chinese remains such as those from Jinniushan and Tianyuan Cave (Shang et al. 2007). Similarly, the African evidence does not, as is often claimed, present "anatomically modern humans" at 150 ka or almost 200 ka. The skulls from Omo Kibish offer some relatively modern features as well as substantially archaic ones;

especially Omo 2 is very robust indeed (McDougall et al. 2005). Their dating, also, is not secure at all, and Omo 2 is a surface find. The much more complete and better dated Herto skull, BOU-VP-16/1, is outside the range of all recent humans in several cranial measurements (White et al. 2003) and is clearly just as archaic as other specimens of the late Middle Pleistocene, in Africa or elsewhere. The lack of "anatomically modern" humans from sub-Saharan Africa prior to the supposed Exodus is glaring: the Border Cave specimens have no stratigraphic context and are thought to be only around 80 ka old; Omo and Dar es Soltan are obviously not sub-Saharan (and the latter is undated), which leaves only the lower jaws of Klasies River Mouth, lacking cranial and postcranial remains. On the other hand, current Australians average a cranial capacity of only 1264 cc (males 1347 cc, females 1181 cc, i.e., well within the range of *Homo erectus*), while their molars average the size of those of Europeans several hundred millennia ago. And yet they are still considerably smaller than those of fossil Australians, such as the large Kow Swamp sample. So while diminution of molars did occur in Australia, supposedly also settled by Eve's progeny, it lags greatly behind that of the rest of the world.

With the lack of African fossils of the African Eve "species," the Eve apostles turned to the Levant for help, and recruited the Mount Carmel finds from Qafzeh Cave and Skhul Shelter as supposed "Moderns." Yet all of these skulls present prominent tori and receding chins, even Qafzeh 9, claimed to be of the most modern appearance. The distinct prognathism of Skhul 9 matches to that of "classic Neanderthals," and the series of teeth from that cave has consistently larger dimensions than typical "Neanderthal" teeth. Even Stringer concedes that this material is "transitional" or intermediate. Besides, supposedly much later "Neanderthal" burials in nearby Tabun Cave as well as the Qafzeh and Skhul material are all associated with the same Mousterian tools, and the datings of all Mount Carmel sites are far from soundly established, with their many discrepancies. The TL dates from Qafzeh, for instance, clash severely with the amino racemization dates (ranging from 33 to 45 ka), and are in any case plagued by inversion: the lower layer (XXII) averages 87.7 ka, the middle layer (XIX) 90.5, while the uppermost (XVII) averages 95.5. Therefore, the claims of 90-ka-old "modern" humans from Mount Carmel, a cornerstone in the Eve notion, are in every respect unsound, and this population is best seen as transitional between robust and gracile forms, from a time when gracilization had commenced elsewhere as well.

This presents an overall picture that is very different from that which the replacement protagonists subscribe to. Their model cannot tolerate such intermediate forms, nor can it allow hybrids, yet in Europe there is a clear continuation of some Neanderthaloid features right up to and into the Holocene. This is demonstrated not only by the Hahnöfersand specimen, but also by others, such as the equally robust "Mesolithic" skull fragment from Drigge, also from northern Germany, which is about 6250 years old (Terberger 1998), and numerous other late specimens previously thought to be of the EUP. They range in age from the "Magdalenian" through the "Neolithic," and even younger. One distinctive "Neanderthal" feature is the shape of the mandibular nerve canal, surrounded by a bony ridge in 53%

of specimens included in this designation. Its occurrence diminishes during the transition period to 44%, but it is still present in today's Europeans, at 6% (Lewin 2005: 196). This feature alone demands the presence of "Neanderthal" genes in Europeans. The process of gracilization has indeed generally continued to the present time: early Mesolithic material is about 10% more robust than modern Europeans. Indeed, Hawks (1997) has estimated that at least 25% of the ancestors of later Upper Paleolithic people would need to be Neanderthals to account for the preservation of Neanderthal autapomorphies observed (see also Frayer 1993, 1998; Frayer et al. 1994).

The second issue emerging from this brief review is that there are now almost no supposedly modern specimens left as possible contenders for attribution to EUP or Aurignacoid industries. The maxilla from Kent's Cavern, United Kingdom (~31 ^{14}C ka BP), and the Romanian remains from Pestera Cioclovina (~29 ^{14}C ka BP) lack secure and diagnostic archeological association. There are, however, numerous "Neanderthal" remains to fill this void. Of particular interest are the most recent, those from Saint Césaire (~36 ka), Arcy-sur-Cure (~34 ka), Zafarraya Cave (~33.4 ka), Máriaremete Upper Cave (~38 ka), Sungir' (~25 ka), Trou de l'Abîme (~33 ka), and Vindija Cave (~28 and ~29 ka).

At the first of these sites, the Neanderthal remains of a burial occur together with clear Châtelperronian artifacts, which until 1979 had been generally assumed to be the work of anatomically modern humans. Arcy-sur-Cure, also in France, yielded numerous ornaments and portable art objects, again from a Châtelperronian. This prompted various convoluted explanations of how these elaborate pendants could have possibly found their way into a "Neanderthal" assemblage (e.g., White 1993; Hublin et al. 1996; a similar argument was used by Karavanic and Smith [1998] in explaining the bone points of Neanderthals in Vindija layer G1). It was contended that the primitive Neanderthals must have scavenged these objects from the camps of "Moderns," as if people lacking the ability to use symbols would have any use for symbolic artifacts. On the other hand, Zafarraya Cave, near Malaga, provides late Mousterian tools (Hublin et al. 1995). The Jankovichian or Trans-Danubian Szeletian (Allsworth-Jones 1986) has yielded three mandibular "Neanderthal" teeth from Máriaremete Upper Cave (Gábori-Csánk 1993). The Streletsian of Sungir' in Russia produced an isolated Neanderthaloid tibia from a triple grave of "Moderns," and the adult male exhibits pronounced supraorbital tori (Bader 1978). Trou de l'Abîme near Couvin in southern Belgium furnished "Neanderthal" remains together with a typical Aurignacian industry, and there can be no question that the Vindija late Neanderthals used EUP tools and technology. Not only has that site supplied some of the most recent "Neanderthals" found so far—and from a site in south-central Europe at that—these are more gracile than Neanderthals of much earlier periods, and they are considered to be transitional by some (Smith and Ranyard 1980; Wolpoff et al. 1981; Frayer et al. 1993; Wolpoff 1999; Smith et al. 2005). Vindija Vi-207 is a mandible of 29,080 ± 400 carbon years BP (OxA-8296), Vindija Vi-208 is a parietal of 28,020 ± 360 carbon years BP (OxA-8295) (Smith et al. 1999). These "late Neanderthals" (or very robust Graciles) exhibit significant reduction in "Neanderthaloid" features such as mid-facial prognathism and supraorbital tori. The related tool assemblage includes even apparent bone fabricators (Ahern et al. 2004).

Ignoring these many significant contradictions to their ideas, the replacement proponents have responded to the recent developments in Germany by contending that the new data bolster their model, because the "Neanderthaloid" Hahnöfersand specimen had been suggested to be a hybrid (Bräuer 1980). In this futile argument they are grasping at straws, instead of admitting the plainly obvious: they have been the victims of the greatest scam in the history of paleoanthropology since the Piltdown affair early in the twentieth century. By hailing each of the very late dates for Neanderthal remains as they appeared in recent years as a confirmation of their prediction that the evidence "effectively precludes any hypothesis of a gradual evolution from Neanderthal to anatomically modern populations within Western Europe itself" (Mellars and Stringer 1989: 8) they have literally argued themselves into a corner. They had strongly contended that "a whole spectrum of radical cultural innovations" (op. cit.) appeared with the beginning of the Aurignacian. But what are the many innovations at 45–40 ka they claim were introduced from Africa? According to them, the people of the Aurignacian are "indistinguishable" from us in terms of cognition, behavior, and cultural potential. Perhaps this is so, but what the evidence now shows is that the period from 45 ka to 28 ka BP has produced dozens of "Neanderthal" specimens, but no securely dated, unambiguously fully modern human remains. This point is reinforced by the occurrence of Neanderthal finds together with EUP lithic traditions at six sites at least, while no "Moderns" have so far been found in clear association with Aurignacian or any other EUP artifacts (Churchill and Smith 2000b). Therefore, one would have thought that the proposition to test is not whether the replacement advocates were correct, but the proposition that the Aurignacian and other Aurignacoid or EUP industries are traditions of "Neanderthals," or of their descendants. Unless that proposition is refuted, we are left with the dictum coined by the African Eve advocates themselves: *that the EUP people, i.e. late 'Neanderthals', from about 45 ka BP on, were of "entirely modern behavior"* (cf. d'Errico 1995).

The "short-range" advocates have apparently failed to grasp the effects of the new data on their embattled hypothesis (e.g., Mellars 2005). There are only three realistic alternatives to account for the "EUP" tool, rock art, and portable art traditions: that they are the work of "Neanderthals"; or of the descendents of "Neanderthals"; or of invading, perhaps genocidal "Moderns." Since there is currently no evidence for the third possibility, and the two others are entirely unacceptable to the mitochondrial Eve advocates because they would refute their hypothesis, one would have thought that they might reconsider. Certainly, the onus is presently on these scholars to present evidence that there were anatomically fully modern humans, free of any "Neanderthaloid" features, in Europe during the first half of their "Upper Paleolithic," i.e., since 45 ka BP. They also need to demonstrate evolutionary continuity in southern Africa, but its complete absence everywhere else. Until they do this, their contentions about human evolution over this period, especially in the European theatre, are contradicted by all available skeletal evidence, and in fact stand refuted.

The African Exodus

Although the notion of replacement arose initially from skeletal evidence (and its misdating) it soon sought support in genetic data. For African Eve advocates, genetic changes in populations represent mass movements of people. Therefore, to render this notion credible, the second phenomenon that would need to be substantiated by hard evidence would be to demonstrate unambiguously that the genetic signatures immediately before and after the claimed replacement are sufficiently different to indicate separate species. Let us be quite clear upfront: no such evidence has been presented. Fragmentary sequences from contaminated remains of robust individuals significantly predating the "replacement" (such as the original August 1856 finds from the Kleine Feldhofer Cave in the Neander valley) show limited differences from present-day populations, which is to be expected. But we have no indication of the genetic distance between Robusts and Graciles of similar ages, say, 30 ka. That, however, is required to show that this distance, and not the distance between specimens of greatly different antiquities, would be great enough to prevent the production of fertile offspring. We already know that individuals living at very different times produced no offspring—they could not breed if they lived in different eras.

However, when we examine the genetic justification of the Eve model (e.g., Cann et al. 1987) we find that it is just as flawed as its skeletal rationalizations (see preliminary notes in Chapter 1). Different research teams have produced different genetic distances in nuclear DNA, i.e., the distances created by allele frequencies that differ between populations (e.g., Vigilant et al. 1991; Barinaga 1992; Ayala 1996; Brookfield 1997). For instance, when the haplotype tree offered by the Vigilant team in 1991 was reanalyzed, its basal branches included non-African origins of human mitochondria. Some geneticists concede that the model rests on untested assumptions; others even oppose it (e.g., Barinaga 1992; Hedges et al. 1992; Maddison et al. 1992; Templeton 1992, 1993, 1996; Brookfield 1997). The various genetic hypotheses about the origins of "Moderns" that have appeared over the past few decades placed the hypothetical split between these and other humans at times ranging from 17 to 889 ka BP. They all depend upon preferred models of human demography, for which no sound data at all are available. This applies to the contentions concerning mitochondrial DNA (African Eve) as much as to those citing Y-chromosomes ("African Adam"; Hammer 1995). The divergence times projected from the diversity found in nuclear DNA, mtDNA, and DNA on the nonrecombining part of the Y-chromosome differ so much that a time regression of any type is extremely problematic. Contamination of mtDNA with paternal DNA has been demonstrated in extant species (Gyllensten et al. 1991; Awadalla et al. 1999; Morris and Lightowlers 2000; Williams 2002), in one recorded case amounting to 90% (Schwartz and Vissing 2002). Not only was the assumption by Cann et al. (1987) about exclusive maternal transference of mitochondria without basis, the constancy of mutation rates affecting mtDNA was also a myth (Rodriguez-Trelles et al. 2001, 2002), because

> molecular time estimates suffer from a methodological handicap, namely, that they are asymmetrically bounded random variables, constrained by a nonelastic boundary at the lower end, but not at the higher end of the distribution. This introduces a bias toward an overestimation of time since divergence, which becomes greater as the length of the molecular sequence and the rate of evolution decrease ... Despite the booming amount of sequence information, molecular timing of evolutionary events has continued to yield conspicuously deeper dates than indicated by the stratigraphic data. Increasingly, the discrepancies between molecular and paleontological estimates are ascribed to deficiencies of the fossil record, while sequence-based time tables gain credit. Yet, we have identified a fundamental flaw of molecular dating methods, which leads to dates that are systematically biased towards substantial overestimation of evolutionary times (Rodriguez-Trelles et al. 2002).

Kidd et al. (1996) have shown that, outside Africa, the elements of which haplotypes are composed largely remain linked in a limited set. Gibbons (1998) observed that by using the new putative "genetic clock," Eve would not be 200 ka old, but only 6000 years. By then the issue had become farcical: the thesis by Cann et al. had not only been based on botched computer modeling, its haplotype trees were fantasies that could not be provided with time depth even if they were real. To render these issues even more complex, the transfer of genetic information is not, as many seem to assume, limited to DNA. For instance, ribonucleic acids associated with the brain's thrombospondin (THBS4 and THBS2) can carry such information (Christopherson et al. 2005; Cáceres et al. 2007), and epigenetic, behavioral, and symbolic inheritance systems need to be considered as well (Jablonka and Lamb 2005), as we shall study in Chapter 5.

The genetic picture in Africa as well as elsewhere has been found to be far more complicated than the Eve proponents ever envisaged. The much-promoted claims that "Neanderthals" were genetically different from modern Europeans, based on very fragmentary DNA sequences, were erroneous, Gutierrez et al. (2002) have shown. Their analysis suggests that the pairwise genetic distance distributions of the two human groups overlap more than claimed, if the high substitution rate variation observed in the mitochondrial D-loop region (Walberg and Clayton 1981; Torroni et al. 1994; Zischler et al. 1995) and lack of an estimation of the parameters of the nucleotide substitution model are taken into account. Moreover, the results presented from museum specimens, especially "Neanderthal" remains, are probably irrelevant. Pruvost et al. (2007) have recently shown that DNA deteriorates rapidly after excavation, up to fifty times as fast as in buried specimens. The various reported "fragmentary DNA sequences" from "Neanderthal" remains stored for up to 150 years need to be considered in that light. A large part, on average 85%, of the genetic material preserved in fossils is lost as a result of treatment by archeologists and storage in museums; therefore, the results disseminated from these specimens and their interpretations may be questioned. More reliable are genetic studies of living populations, which have shown that both Europeans and Africans have retained significant alleles from multiple populations of Robusts (Hardy et al. 2005; Garrigan et al. 2005; cf. Templeton 2005). In fact, the Neanderthal genome seems to include an excess of "human"-derived single nucleotide polymorphisms (Green et al. 2006).

Recent genetic analyses confirm not only that "Neanderthal" genes persist in recent Europeans, Asians, and even Papuans (Green et al. 2010), but also that "it seems Neandertals interbred with the ancestors of Europeans and Asians, but not with the ancestors of Africans" (Gibbons 2010; cf. Krings et al. 1997). In the words of Green et al. "[g]iven that the OOA alleles occur at a frequency of much less than 50% in non-Africans (average of 13%, and all less than 30%), the fact that the candidate regions match the Neandertals in 10 of 12 cases ($P = 0.019$) suggests that they largely derive from Neandertals". Thus, the African Eve model has become an absurdity: it is precisely Africans who had the least contact with Europeans.

Relethford (2002) has detected drastic spatiotemporal changes in the genetic profiles of three recent Chinese populations, negating the idea of regional genetic homogeneity. He found that the Linzi population of 2500 years BP is genetically more similar to present-day Europeans than to present-day eastern Asians. This refutes the idea that regional comparisons of DNA can establish affinity or its absence. Assumptions about a neutral mutation rate and a constant effective population size are completely unwarranted, and yet these variables determine the outcomes of all the genetic calculations. For instance, if the same divergence rate as one such model assumes (2–4% base substitutions per million years) is applied to the human–chimpanzee genetic distance, it yields a divergence point of 2.1–2.7 million years, which can be considered unambiguously false. Nei (1987) suggests a much slower rate, 0.71% per million years (now abandoned by him), according to which the human–chimpanzee separation would have occurred 6.6 million years ago, which is close to the estimate from nuclear DNA hybridization data of 6.3 million years. It also appears to be close to what the fossil record seems to indicate. However, this would produce a divergence of "Moderns" at 850 ka BP, over four times as long ago as the favored models, and eight times as long ago as the earliest fossils of *purported* Moderns ever found (though both their dating and modernity are controversial, as noted above). To explain away the perplexingly late split of the "Moderns", some of the short-range geneticists have even resorted to suggesting mtDNA transfer between "proto-humans" (e.g., australopithecines) and proto-chimpanzees (i.e., species presumably separated by millions of years of evolution), while at the same time excluding such a possibility for recent robust and gracile populations that are of the same species (Hasegawa et al. 1985).

Molecular archeology, the analysis and interpretation of paleogenetic, ancient DNA (Pääbo 1989), remains an experimental method and those who overinterpret its results tend to overlook its limitations. Initial results were obtained from a quagga (Higuchi et al. 1984), an Egyptian mummy (Pääbo 1985), a moa (Cooper et al. 1992), and a cave bear (Noonan et al. 2005), before the genome of *Homo sapiens neanderthalensis* was tackled (Green et al. 2006). But paleogenetics poses challenges that differ significantly from in vivo studies, because DNA suffers both mechanical and chemical degradation through time and there are high sequencing error and template damage rates (Pääbo et al. 2004; Pruvost et al. 2005, 2007; Orlando and Hänni 2008). It is certainly easier to template modern DNA than ancient DNA. Results of the polymerase chain reaction (PCR) amplifications, performed by clonage, need to be repeated and three negative controls have to be added to safely detect contamination. Then there is the potential, particularly in moist conditions, of hydrolytic cleavage of phosphodiester bonds between phosphate and sugar (Jolivet and Henry 1994: 180). Similarly, sugars and amino groups in proteins and nucleic acids, caused by condensation, can react and lead to errors during PCR. Deamination of cytosine in xanthine, guanine, and uracil, or adenine in hypoxanthene can occur, involving the incorporation of nucleotide in the process of PCR amplification. The issues of base substitution (Lindhal and Nyberg 1972) and fragmentation of DNA (Golenberg et al. 1996) have long been known, and the point is demonstrated, for instance, by the erroneous results obtained from the DNA of insects embedded in amber (Gutierrez and Marin 1998). Other problems with interpreting or conducting analyses of paleogenetic materials are alterations or distortions through the adsorption of DNA by a mineral matrix, its chemical rearrangement, microbial or lysosomal enzymes degradation, and lesions by free radicals and oxidation (Geigl 2002; Carlier et al. 2007, 2008). These scientific qualifications are generally unheeded in the archeological folklore established around the "authoritative" DNA data, in much the same way as archeologists usually fail to heed the reservations of scientists concerning datings of rock art (Bednarik 1996, 2002; Watchman 1999), or most other archeometric data furnished by scientists. Such results are always grossly simplified, misinterpreted, and overinterpreted, and then embedded in the mythology of mainstream archeology. In the case of paleogenetic data, they have been eagerly seized by one or another school of thought to support its case or discredit that of the opponents. Yet archeologists make little concerted effort to appreciate the reservations scientists have.

For instance, there are considerable complexities concerning the accumulation of base substitutions, or mutations, that are not even relevant to natural selection. The mechanisms governing DNA mutation rates, which are so central to the archeological claims involving genetics, are not at all well understood. Those mutations that have no selection consequences, "neutral" mutations, are also reflected in DNA mutation rates, which can be estimated by comparing neutrally evolving sequences in species that share a common ancestor. Sequences that are high in pairs of the bases C and G (CpGs) have been positively correlated with mutation rate. However, the chemical modification of CpGs makes them prone to mutation themselves, and with time they are eliminated from neutrally evolving sequences. Walser and Furano (2010) have taken advantage of this property to investigate the role of CpGs on the mutation rate of non-CpG DNA by comparing "old" and "young" sequences. They found that CpGs are not only promoting mutations, but also influencing how the non-CpG sequences around them are being mutated. In determining the neutral non-CpG mutation rate as a function of CpG content they compared sequence divergence of thousands of pairs of neutrally evolving chimpanzee and human orthologs that differ primarily in CpG content. Both mutation rate and mutational processes are contingent on the local CpG content.

Protagonists of the replacement or short-range hypothesis often cite possible genetic bottlenecks to contrive explanations for inherent weaknesses of their model. But in combining the model of a population bottleneck with that of an endemic population we also need to remember that genetic bottlenecks tend to reduce fitness in the population (Bryant et al. 1986; Berger et al. 2008), rather than bring about the population's supremacy (cf. Hawks et al. 2000), as proposed for Eve's progeny. Another genetic model (Pennisi 1999) has modern humans evolving from two discrete populations, one resulting in modern Africans, the other in non-Africans. Templeton (2002) has contradicted the replacement hypothesis genetically. Using ten different haplotype trees (MtDNA, Y-chromosomal DNA, two X-linked regions and six autosomal regions), he suggested that following an initial exodus from Africa at about 1.7 million years ago, there were at least two subsequent major expansions out of Africa. One occurred at 840–420 ka ago, the second at 150–80 ka ago. The genetic data also show ubiquity of genetic interchange or interbreeding between human populations *throughout the 1.7 million years*, as had been predicted by the multiregional hypothesis since Weidenreich (1946). In response to Templeton's comprehensive refutation of the Eve model, Cann (2002) made no attempt to argue against his proposals of long-term, multiregional evolution.

It is also of concern that the first colonization dates assumed by the geneticists supporting the Eve model are mostly false (see e.g., Cann et al. 1987), and these researchers admitted from the beginning of their involvement that their base-pair substitution rates were based on the (almost certainly false) assumption of single colonization events. It has long been known that there were probably multiple settlement events in most colonizations, which in the case of Australia is indisputable (Bednarik and Kuckenburg 1999). The dingo, introduced in the mid-Holocene, could not have arrived unassisted. The lineage of the earliest known "anatomically modern" remains in Australia, Lake Mungo 3, has been shown to have probably diverged before the most recent common ancestor of contemporary human mitochondrial genomes (Adcock et al. 2001). In the absence of any reliability of the proposed rates of nucleotide changes and the many variables to be accounted for effectively, the contentions by the replacement advocates are essentially unsupported, and nucleotide recombination renders their views fully redundant (Strauss 1999).

When the same "genetic clock" used in all this is applied to dogs and suggests that the split between wolves and dogs occurred 135 ka ago, archeologists reject it on the basis that there is no paleontological evidence for dogs prior to about 15 ka BP (Napierala and Uerpmann 2010; but see Germonpré et al. 2009 for a claim of c. 31 ka). In other words, the weak theory that effectively provides the only remaining basis for the replacement scenario is rejected when applied to another species. The scenario of genetic isolation, long enough to render Eve's progeny unable to interbreed with any other humans, is another unsupportable short-range notion. Interfertility yielding viably breeding offspring occurs between many species (e.g., in wolf, coyote, and dog; in several species of deer; in mallards and ducks) and can even yield viable subspecies.

Instead of unambiguously showing that "anatomically modern humans" (whatever that ethnocentric term is intended to mean) originate in one region, sub-Saharan Africa, all the available genetic data suggest that gene flow occurred in Old World hominins throughout much of recent human evolution (Templeton 1996, 2002), which is also strongly suggested by all available empirical evidence, both paleoanthropological and archeological. For instance, the evidence that *Homo sapiens neanderthalensis* managed to live and subsist at the Arctic Circle, in temperatures that would at times have been below –40°C (Schulz 2002; Schulz et al. 2002; see also Pavlov et al. 2001), easily dispatches the notion that there were great expanses of habitable land in Europe that remained unoccupied by humans. The Finnish evidence, dating back 135 ka BP, suggests that these innovative people coped with extreme climatic conditions by the advent of the Late Pleistocene, and that the demographic modeling of Pleistocene archeologists (e.g., Gamble 1999) must be largely false. If human groups on the margins were forced into regions of truly appalling living conditions we can safely assume the presence of largely continuous populations in much of the Old World, and by 50 ka even in Australia.

That implies that *Homo sapiens sapiens* must have evolved as a single extended breeding unit across much or most of the region once occupied by robust *sapiens* hominins, from southern Africa to eastern Asia and Australia. Genetic drift, introgressive hybridization (Anderson 1949), and episodic genetic isolation during climatically unfavorable events (e.g., the Campagnian Ignimbrite event, or the Heinrich Event 4; Barberi et al. 1978; Fedele et al. 2002, 2003; Fedele and Giaccio 2007) rather than mass migration probably account for the mosaic of hominin forms we can observe through time. Reticular introgression tends to increase at times of ecological stress. Previously deleterious variants, be they mutation- or introgression-derived, may become adaptive. Episodic sharp reductions in gene pool size are the most effective factor in the acceleration of phylogenetic change in a population, particularly if they are combined with genetic drift or introgression across contiguous populations subjected to demographic adjustments.

It is then unnecessary to resort to an explanation via mass migration and complete replacement by an intrusive population. The many archaic or robust *sapiens* populations of the Old World existed in various climatic and ecological environments, therefore would have had a much higher number of adaptive mutations, and would have genetically overwhelmed a numerically smaller intrusive population bringing with it a much smaller number of adaptive alleles. Moreover, since the technology of any contemporary robust and gracile populations were always evenly matched (as we shall see below), the notion that some African super-humans overwhelmed an extremely well adapted and acclimatized resident population in Europe and everywhere else is demographically most unlikely. Such replacement could have been through introduced diseases, it has been argued; but that contention, surely, cuts both ways. Since we need to assume, as a credible null-hypothesis, that all habitable parts of Europe were occupied by robust tribes 40 ka ago, we would have a scenario of cold-adapted, physically much stronger Robusts being attacked at the periphery by numerically inferior, naked people from the tropics with their thin skulls, wielding the same kinds of Mode 3 weapons. One does not need to be a military strategist to see who would have overwhelmed whom, both physically and genetically. Moreover, in any massive invasion, much of the mitochondrial DNA of the ensuing population turns out to be that of the vanquished, whose females the victors absorb.

Ultimately, demographic genetics, that is, allele drift based on generational mating site distance, easily accounts for archeologically observed population changes (Harpending et al. 1998). A distance of merely 50 km per generation is most reasonable for such highly mobile populations, and suffices to explain the travel of genes over 10,000 km in as few as 200 generations. In reality, generational mating site distances of some hundreds of kilometers seem to be perfectly reasonable. Yet the enormous time scale available for the development of "Moderns" amounts to perhaps 2000 generations. The notion of invading Africans is as likely to be valid as the account of Noah's Ark: the more one thinks about the logistics and demography, the more absurd it tends to appear—except for the believers.

Tools and Cultures

We have seen that the Eve model derives to some extent from a series of false dates of fossil hominin remains and that Graciles emerged not suddenly, but gradually over many millennia. Next we considered the claims of genetic support, and we found none: invented coefficients of all crucial variables, false assumptions, and computer bungles characterize this aspect. Already the replacement hypothesis looks invalidated, yet we have not even reviewed some of the most debilitating factors. In Chapter 1 we considered the epistemology of taxonomies of Pleistocene archeology and it seemed that the cultures and their sequence are based on invented stone tool categories, rather than on authentic cultural variables. So to trust them invites circular reasoning, such as: these purported cultures must have some validity, even if they are only based on tool types, because similar combinations do occur repeatedly and seem to have discrete geographical distributions. But the tool types were derived from their occurrence in certain combinations, and they as well as the cultures are invented constructs. The process of validation is always one of confirmation, because whenever it seems to fail, we confirm it by contrasting the failed occurrences and naming them other tool types, other combinations of tool types, other cultures. This is an unscientific procedure, in the sense that it implicitly rejects testing by simply finding excuses for refuting evidence.

Despite these concerns about the taxonomies, and for the sake of the argument, we will now pretend that these entities do have real existence in the period we are concerned with, roughly from 45 to 25 ka ago. We will determine whether the empirical pronouncements based on these taxonomies, as considered below, will fail when tested against the predictions and claims of the replacement model.

The EUP industries of Eurasia first appear fairly simultaneously between 45 ka and 40 ka BP, or even earlier, at widely dispersed locations from Spain to Siberia (e.g., Makarovo 4/6, Kara Bom). Senftenberg, a clearly Upper Paleolithic (Mode 4) blade industry in the middle of Europe (Gravette point, keeled Aurignacian scraper) has even been dated to 48,300 ± 2000 (GRO-1217) or, if we are to consider a still earlier date, >54,000 years BP (GRO-1771) (Felgenhauer 1959: 60). The Aurignacian of El Castillo level 18, in Spain, seems to commence well before 40 ka ago (Cabrera Valdés and Bischoff 1989; carbon dates of 40,000 ± 2100, 38,500 ± 1800, 37,700 ± 1800 BP). At Abric Romani, the lowest AMS dates from the Aurignacian average 37 ka BP, but the probably more relevant uranium-series dates point to a sidereal age of 43 ka BP (Bischoff et al. 1994). At El Pendo, the Lower Périgordian (i.e., Châtelperronian) industry, attributed to "Neanderthals" in France, overlies two Early Aurignacian levels (González Echegaray et al. 1980), a stratigraphic pattern also observed in France, for example, at Roc de Combe (Bordes and Labrot 1967) and La Piage (Champagne and Espitalié 1981). The Châtelperronian at Morín Cave has been dated to about 36,950 carbon-years BP, an antiquity similar to that of the same tradition at French sites (generally 37–33 ka BP). The most recent Middle Paleolithic (Mode 3) occupation known in Spain, however, is at Abric Agut. According to both radiocarbon and U-series dating, it occurred only 13–8 ka BP, i.e., straddling the Pleistocene–Holocene interface (Vaquero et al. 2002). Like many other finds, it shows how illusory the separation of the Middle and Upper Paleolithic cultures is (Bednarik 1995b).

The Iberian pattern of a mosaic and gradually decreasing component of Mode 3 technology in regional EUP lithic industries applies through much of Europe. In southern Italy, variants such as the Uluzzian (Palma di Cesnola 1976, 1989), the Uluzzo-Aurignacian, and the Proto-Aurignacian (43–33 ka BP) have been reported (Kuhn and Bietti 2000; Kuhn and Stiner 2001). The Olschewian of the Alpine region, another Aurignacoid tradition (42–35 ka BP), developed from the final Mousterian (Abel 1931; Andrist et al. 1964; Bächler 1940; Bayer 1924, 1928, 1929a, b, 1930; Bednarik 1993, 2007; Bégouën and Breuil 1958; Brodar 1957; Cramer 1941; Ehrenberg 1951, 1953a, b, 1954, 1956, 1957, 1958, 1959, 1962, 1970; Kurtén 1968: 127; Kyrle 1931; Malez 1956, 1959, 1965; Mottl 1950; Rabeder et al. 2000; Rakovec 1967; Stehlin and Dubois 1916; Trimmel 1950; Tschumi 1949; Vértes 1951, 1955, 1959, 1965; Zotz 1939, 1944, 1951). Further east this mosaic includes the Bachokirian of the Pontic region (>43 ka BP), the Bohunician of east-central Europe (Svoboda 1990, 1993: 44–38 ka BP), and various traditions of the Russian Plains. The latter comprise major concentrations of sites in the Prut-Dniester basin and on the middle Don. Some of these industries, such as the Streletsian, Gorodtsovian, and Bryynzenian derived clearly from Mousteroid technologies, whereas the Spitzinian or Telmanian are free of Mode 3 bifaces (Anikovich 2005). In parts of Russia, such as regions of the Don River, the Crimea, and northern Caucasus, Mode 3 technologies (Mousterian and Eastern Micoquian) continue alongside intermediate and Mode 4 ones and the gradual development from one into the other can be observed at many individual sites. The coexistence of seven accepted tool traditions between 36 ka and 28 ka BP has been reported from the region: the Mousterian, Micoquian, Spitzinian, Streletsian, Gorodtsovian, Eastern Szeletian, and Aurignacian (Krems-Dufour variant). The rich mosaic of "EUP cultures" began before 40 ka BP on the Russian Plain and ended only 24–23 ka BP. In the Crimea, the Middle Paleolithic is thought to have ended only between 20 and 18 ka BP. Elsewhere in the region, the introduction of a first fully developed Upper Paleolithic tradition (the Kostenkian) appears about 24 ka at the Kostenki-Borshevo site complex.

A succession of traditions connecting Middle Paleolithic biface technocomplexes, including the late Eastern Micoquian, with typical late Paleolithic ones, continue through the Szeletian of eastern Europe (Allsworth-Jones 1986; 43–35 ka BP), the Jankovician of Hungary; and the Altmühlian (c. 38 ka BP), Lincombian (38 ka BP), and Jerzmanovician (38–6 ka BP) further north. Similarly, the gradual development from the Middle Paleolithic at 48 ka BP (with "Neanderthal" footprints of small children) to the Upper Paleolithic is clearly documented in Theopetra Cave, Greece (Kyparissi-Apostolika 2000; Facorellis et al. 2001). These and other cases of "intermediate" industries or gradual changes all demonstrate the continuity between Mode 3 and Mode 4 technocomplexes in many parts of Europe, but most especially in the east and southeast, the logical entry point of the presumed African invaders. A degree of regionalization precedes this period even in the late Mousterian (Kozłowski 1990; Stiner 1994; Kuhn 1995; Gamble 1999; Riel-Salvatore and Clark 2001), marked by both miniaturization and increasing use of blades, by improved hafting and the use of backed or blunted-back retouch, apparently heralding subsequent developments. German Mode 3 sites have produced backed microliths and evidence of the use of birch resin, and replication experiments suggest that the technology involved in preparing this resin is exceedingly complex. The artificial dichotomy between Middle and Upper Paleolithic materials has thus only served to overemphasize gradual changes in technology (Fedele et al. 2003), at the expense of veracity.

The specious separation of Mode 3 and Mode 4 technologies has even less currency in Africa (e.g., the Howieson's Poort tradition with its microliths, or the Amudian), India (Bednarik 1994; Bednarik et al. 2005) or China (Gao and Norton 2002). In Australia the Mode 3 traditions continue until well into the Holocene, and in Tasmania until the arrival of the British, just over two centuries ago. Indeed, at Tasmania a society with a Mode 3 technology was ethnographically observed.

Instead of a sudden change of technology in Europe at any time during the period from 45 to 25 ka ago, what we do observe is a complex mosaic of regional traditions which, in general, exhibit a gradual change of several variables, such as tool size, knapping method, retouch, and reuse. This suggests in all cases *in-situ* evolution of cultures, rather than the effects of an intrusive tradition. It mirrors precisely the patterns already documented above, in the development in human morphology. Moreover, not one of the more than twenty perceived EUP stone tool traditions of Europe derives from Africa or the Levant. On the contrary, Aurignacoid or similar traditions arrived in the Levant long after they first arose elsewhere in Eurasia, so they were clearly not introduced through this presumed corridor. If these technologies had been imported from sub-Saharan Africa, one would expect a trail of their traces through northern Africa. Yet right across the north of that continent, the Middle Stone Age continued up to 20 ka ago, i.e., at least for 20,000 years after the introduction of Mode 4 technologies across Eurasia. The advocates of the Eve model have steadfastly ignored this glaring inconsistency. Nor have they ever explained where the African or Levantine precedents of the Upper Paleolithic art traditions are to be found, if these African invaders were their carriers as claimed. There is no trace of such evidence, in fact there is almost no proven Pleistocene rock art currently known from Africa, the exception being very early petroglyphs from the Kalahari (Beaumont and Bednarik 2010). Similarly, the only demonstrated mobiliary art from Africa, found in Namibia, is merely 26 ka or so old (and is not of the MSA, as often claimed, but of the middle LSA). The state of available information from the Levant or Arabia indicates much the same along the route the Africans are supposed to have taken to Europe.

This raises yet another matter consistently avoided by the proponents of replacement. Wherever robust and more gracile forms of humans apparently coexisted locally (as was often the case for tens of millennia), be it in the Levant, in Australia or in any part of Europe, they are always thought to have shared quite similar if not identical cultures, technologies, even ornaments. The simplistic notion that one can trace ethnic differences through tool assemblages is, therefore, unlikely to be helpful in understanding the cultural dynamics of this period; it is yet another fundamental fallacy of archeology. Not only does this observation render the idea of technological or cognitive superiority of the Graciles without support, it reminds us of the tendency of the Eve advocates to explain away any evidence contradicting their version of events. For instance, when sophisticated items of decoration are found together with robust human remains, "they must have been scavenged" from their much more developed gracile betters, or they must "indicate a running ahead of time". When the genetic sophistries propping up the Eve notion contradict the archeological view of the dog's ancestry the finding is rejected. Accommodative modes of thinking of this kind can of course explain away any evidence, but they are complete anathema to scientific practice. The question then becomes whether we are engaged in a search for truth, or whether the purpose of our efforts is to confirm what we already believe to be the case. This is where science, as defined in Chapter 1, truly has to part ways with orthodox archeology, and where the operation of the latter as a belief system becomes apparent.

Until 1979, the Châtelperronian had been considered to be the work of "Moderns," and after its "Neanderthal" makers were recognized, it became essential to explain the presence of a suite of extraordinarily developed decorative objects (Fig. 2.2). The notion of the Châtelperronian being a tool tradition of gracile people can be traced back to Dorothy Garrod, the controversial archeologist who tried to salt the Glozel site on 8 November 1927 in order to discredit its discoverer, the nonarcheologist Émile Fradin. She believed that the "Upper Paleolithic cultures" were the result of a series of invasions into Europe. Since 1979 it has been conceded that the Châtelperronian refers to a technocomplex of robust people. This chapter has shown that history will need to repeat itself, and this time there will be no explaining away of the empirical evidence. Last time around Eve's apostles were able to attribute the incriminating proof to someone else: those smart Aurignacians, who were so very much like ourselves, and so very utterly different from those brutish Neanderthals (how did they ever manage to make Châtelperronian stone tools?). This time checkmate is looming, because it now appears that those smart Aurignacians were also "Neanderthals." In fact a most ignominious defeat now stares the replacement archeologists in the face. Not only the Aurignacian tools, but also those of all other EUP industries are in all probability the work of Robusts:

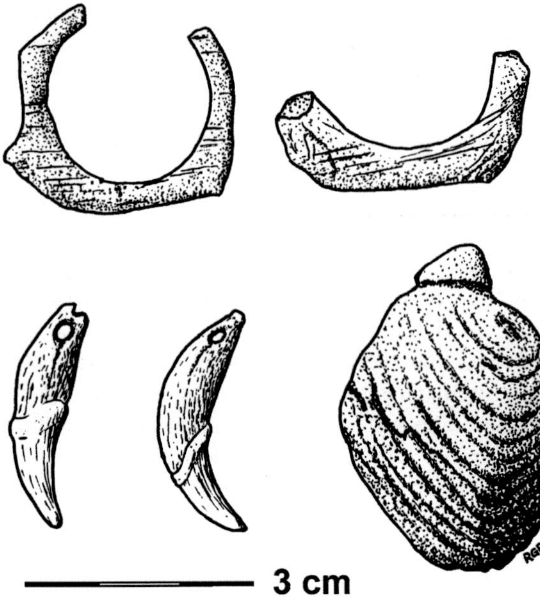

Fig. 2.2 Two ivory ring fragments, two perforated animal canines, and a fossil shell with an artificial groove for attachment. Châtelperronian, Grotte du Renne, Arcy-sur-Cure, France. These objects were used, and almost certainly made, by "Neanderthals"

Neanderthals or their direct descendants. Worse still, modern Europeans themselves, the crown of creation, God's redeeming gift to this world, seem to descend from these Neanderthals. The humiliation is not just about descending from these brutes, but about having to concede that, for some decades, the Eve supporters have led the study of hominin origins on a monumental wild-goose chase.

They will no doubt argue that every possible avenue needed to be explored, and that research will inherently and unavoidably lead to some dead ends as well as viable solutions. However, this particular impasse was not necessary. If we check through the data and rationalizations of this chapter, it becomes clear that most of this information was available before 1987, the time of the rise of the replacement hypothesis to dogma. Of the five conditions listed at the beginning of this chapter as providing the support needed by this hypothesis, four have now been decisively shown not to have been satisfied, and a rational review would have yielded the same conclusion more than 20 years ago. It is clear that the paleoanthropological record of Europe has not produced evidence of a sharp separation between robust and gracile specimens, but instead points to a complex mosaic suggesting some process of speciation. It is equally clear that no evidence has been presented that, at any time in the final third of the Late Pleistocene, contemporary populations exhibited genetic signatures that might indicate the presence of separate species. Nor was there a sudden appearance of pronounced gracility at any point in time. Europeans about 10 ka ago were on average 10% more robust than today (large molars, more robust crania and skeletons). That trend of the average continues further back until we arrive at typical "Neanderthals." We have also found that the same gradual development applies to the technologies. In fact the development of technology from about 30 to

10 ka is clearly greater than that from 50 to 30 ka ago, and all of it is comparatively gradual. Finally, there is no archeological evidence along the presumed route taken by the mythical African invaders, of an intrusive technology or culture that could have arisen in sub-Saharan Africa. Thus, four crucial tests applied to the Eve model have led to its sound rebuttal, and the fifth should not even be required. The "big bang of consciousness," the "creative explosion" marking the arrival of the intrusive population and technology that we cannot find any evidence for on the genetic, skeletal or technological record still remains to be considered. It is very central to the subject of this book, so we will consider it more exhaustively. Suffice it to say, this perspective will best illuminate the follies of the replacement ideas.

Before we move on to examine these more important factors in how and why we became human, we need to consider one more epistemological aspect of the replacement hypothesis: the question of timing. If the Robusts of Europe were overwhelmed by an intrusive population of a different species, that invasion must have begun at a specific point in time. That time would be marked by the first occurrence of the superior technological manifestations that rendered the invasion successful, and by the first evidence of the purported symbolic revolution driving the invasion. But whatever time slot we choose for the event, it is squarely contradicted by most of the empirical data. The "Upper Paleolithic" technology first appears by at least 45 ka ago, but there is no sign of even remotely gracile humans. If we take the advent of Franco-Cantabrian rock art and mobiliary art as the first proof of Moderns, 33 and 40 ka respectively might be realistic estimates, but it is contradicted by technology, and by the presumed identity of the artists as Neanderthaloid. Alternatively, we could set the date by the first anatomically modern skeletal remains, but there is no agreement on their identification, and there is in any case a gradual introduction. Timing it by the end of the fossils we choose to include with the "Neanderthals," or the stone tools we call Middle Paleolithic, would be another option, but the first end 28 ka ago, while the second marker could be set anywhere from 40 to 10 ka ago in Europe, and even much later elsewhere. Not surprisingly, there is no consensus on the timing of the invasion, or even on how it ought to be identified. But without pinpointing this event, by whatever evidence, how can the replacement model have any credibility whatsoever? Unless we have evidence that the first Graciles were genetically so different from the Robusts of their time that they could not produce fertile offspring with them, we have in fact a hypothesis that should never have gained any currency.

Let us now see how this unfounded hypothesis that fails every simple test fares when we examine the relevant evidence that relates to *culture* rather than some other variable.

References

Abel, O. 1931. Das Lebensbild der eiszeitlichen Tierwelt der Drachenhöhle bei Mixnitz. In O. Abel and G. Kyrle (eds), *Die Drachenhöhle bei Mixnitz*, pp. 885–920. Speläologische Monographien, Vols. 7–9, Vienna.

Adcock, G. J., E. S. Dennis, S. Easteal, G. A. Huttley, L. S. Jermiin, W. J. Peacock and A. Thorne 2001. Mitochondrial DNA sequences in ancient Australians: implications for modern human origins. *Proceedings of the National Academy of Sciences of the United States of America* 98(2): 537–542.

Ahern, J. C. M., I. Karavanic, M. Paunović, I. Janković and F. H. Smith 2004. New discoveries and interpretations of fossil hominids and artifacts from Vindija Cave, Croatia. *Journal of Human Evolution* 46: 25–65.

Allsworth-Jones, P. L. 1986. The Szeletian: main trends, recent results, and problems for resolution. In M. Day, R. Foley and W. Rukang (eds), *The Pleistocene perspective*, pp. 1–25. World Archaeological Congress, Southampton 1986. Allen and Unwin, London.

Anderson, E. 1949. *Introgressive hybridization*. Wiley, New York, NY.

Andrist, D., W. Flähiger and A. Andrist 1964. Das Simmental zur Steinzeit. *Acta Bernensia* 3: 1–46.

Anikovich, M. 2005. Early upper paleolithic cultures of Eastern Europe. *Journal of World Prehistory* 6(2): 205–245.

Asmus, G. 1964. Kritische Bemerkungen und neue Gesichtspunkte zur jungpaläolithischen Bestattung von Combe-Capelle, Périgord. *Eiszeitalter und Gegenwart* 15: 181–186.

Awadalla, P., A. Eyre-Walker and J. Maynard Smith 1999. Linkage disequilibrium and recombination in hominid mitochondrial DNA. *Science* 286: 2524–2525.

Ayala, F. J. 1996. Response to Templeton. *Science* 272: 1363–1364.

Bächler, E. 1940. *Das alpine Paläolithikum der Schweiz*. Monographien zur Ur- und Frühgeschichte der Schweiz, Vol. 2, Basle.

Bader, O. N. 1978. *Sungir: Verkhnepaleoliticheskaya stoyanka*. Izdatel'stvo 'Nauka', Moscow.

Barberi, F., F. Innocenti, L. Lirer, R. Munno, T. S. Pescatore and R. Santacroce 1978. The Campanian Ignimbrite: a major prehistoric eruption in the Neapolitan area (Italy). *Bulletin of Volcanology* 41: 10–22.

Barinaga, M. 1992. 'African Eve' backers beat a retreat. *Science* 255: 686–687.

Bayer, J. 1924. Die geologische und archäologische Stellung des Hochgebirgspaläolithikums der Schweiz. *Die Eiszeit* 1: 59–65.

Bayer, J. 1928. Das zeitliche und kulturelle Verhältnis zwischen den Kulturen des Schmalklingenkulturkreises während des Diluviums in Europa. *Die Eiszeit* 5: 9–23.

Bayer, J. 1929a. Die Olschewakultur. *Eiszeit und Urgeschichte* 6: 83–100.

Bayer, J. 1929b. Wildkirchlikultur. *Eiszeit und Urgeschichte* 6: 142.

Bayer, J. 1930. Hat das Hochgebirgspaläolithikum der Schweiz Knochenwerkzeuge geliefert? *Eiszeit und Urgeschichte* 7: 139–140.

Beaumont, P. and R. G. Bednarik 2010. Pleistocene rock art from Africa. Paper presented to IFRAO Congress 2010, Foix.

Bednarik, R. G. 1993. Wall markings of the cave bear. *Studies in Speleology* 9: 51–70.

Bednarik, R. G. 1994. The Pleistocene art of Asia. *Journal of World Prehistory* 8(4): 351–375.

Bednarik, R. G. 1995a. Concept-mediated marking in the Lower Palaeolithic. *Current Anthropology* 36(4): 605–634.

Bednarik, R. G. 1995b. Traces of cultural continuity in Middle and Upper Palaeolithic material evidence. *Origini* 18: 47–67.

Bednarik, R. G. 1996. Only time will tell: a review of the methodology of direct rock art dating. *Archaeometry* 38(1): 1–13.

Bednarik, R. G. 2002. The dating of rock art: a critique. *Journal of Archaeological Science* 29(11): 1213–1233.

Bednarik, R. G. 2006. The cave art of Mladeč Cave, Czech Republic. *Rock Art Research* 23: 207–216.

Bednarik, R. G. 2007. Antiquity and authorship of the Chauvet rock art. *Rock Art Research* 24: 21–34.

Bednarik, R. G. and M. Kuckenburg 1999. *Nale Tasih: eine Floßfahrt in die Steinzeit*. Jan Thorbecke, Stuttgart.

Bednarik, R. G., G. Kumar, A. Watchman and R. G. Roberts 2005. Preliminary results of the EIP Project. *Rock Art Research* 22: 147–197.

Bégouën, H. and H. Breuil 1958. *Les cavernes du Volp, Trois-Frères, Tuc d'Audoubert*. Arts et métiers graphiques, Paris.

Berger, L. R., S. E. Churchill, B. De Klerk and R. L. Quinn 2008. Small-bodied humans from Palau, Micronesia. *PLoS One* 3(3): e1780. DOI: 10.1371/journal.pone.0001780.

Bischoff, J. L., K. R. Ludwig, J. F. Garcia, E. Carbonell, M. Vaquero, T. W. Stafford and A. J. T. Jull 1994. Dating of the basal Aurignacian sandwich at Abric Romani (Catalunya, Spain) by radiocarbon and uranium series. *Journal of Archaeological Science* 21: 541–551.

Bordes, F. and J. Labrot 1967. La stratigraphie du gisement de Roc de Combe (Lot) et ses implications. *Bulletin de la Société Préhistorique Française* 64: 15–28.

Bräuer, G. 1980. Die morphologischen Affinitäten des jungpleistozänen Strinbeins aus dem Elbmündungsgebiet bei Hahnöfersand. *Zeitschrift für Morphologie und Anthropologie* 71: 1–42.

Bräuer, G. 1981. New evidence of the transitional period between Neanderthal and modern man. *Journal of Human Evolution* 10: 467–474.

Bräuer, G. 1984a. A craniological approach to the origin of anatomically modern *Homo sapiens* in Africa and implications for the appearance of modern Europeans. In F. H. Smith and F. Spencer (eds), *The origins of modern humans: a world survey of the fossil evidence*, pp. 327–410. Alan R. Liss, New York, NY.

Bräuer, G. 1984b. Präsapiens-Hypothese oder Afro-europäische Sapiens-Hypothese? *Zeitschrift für Morphologie und Anthropologie* 75: 1–25.

Bräuer, G. 1989. The evolution of modern humans: a comparison of the African and non-African evidence. In P. Mellars and C. Stringer (eds), *The human revolution: behavioural and biological perspectives on the origins of modern humans*, pp. 123–154. Edinburgh University Press, Edinburgh.

Brodar, S. 1957. Zur Frage der Höhlenbärenjagd und des Höhlenbärenkults in den paläolithischen Fundstellen Jugoslawiens. *Quartär* 9: 147–159.

Brookfield, J. F. Y. 1997. Importance of ancestral DNA ages. *Nature* 388: 134.

Bryant, E. H., S. A. McComas and L. M. Combs 1986. The effect of an experimental bottleneck on quantitative genetic variation in the housefly. *Genetics* 114: 1191–1211.

Cabrera Valdés, V. and J. Bischoff 1989. Accelerator ^{14}C dates for early Upper Palaeolithic (Basal Aurignacian) at El Castillo Cave (Spain). *Journal of Archaeological Science* 16: 577–584.

Cáceres, M., C. Suwyn, M. Maddox, J. W. Thomas and T. M. Preuss 2007. Increased cortical expression of two synaptogenic thrombospondins in human brain evolution. *Cerebral Cortex* 17: 2312–2321.

Cann, R. L. 2002. Tangled genetic routes. *Nature* 416: 32–33.

Cann, R. L., M. Stoneking and A. C. Wilson 1987. Mitochondrial DNA and human evolution. *Nature* 325: 31–36.

Carlier, L. 2008. Caractérisation de la région 51-160 de la protéine KIN17 humaine par RMN et modélisation moléculaire, les dommages de l'ADN, 2000–2008.

Carlier, L., J. Couprie, A. le Maire, L. Guilhaudis, I. Milazzo, M. Gondry, D. Davoust, B. Gilquin and S. Zinn-Justin 2007. Solution structure of the region 51-160 of human KIN17 reveals an atypical winged helix domain. *Protein Science* 16: 2750–2755.

Champagne, F. and R. Espitalié 1981. *La Piage, site préhistorique sur Lot*. Mémoires de la Société Préhistorique Française, No. 15.

Christopherson, K., E. Ullian, C. Stokes, C. Mullowney, J. Hill, A. Agah, J. Lawler, D. Mosher, P. Brownstein and B. Barres 2005. Thrombospondins are astrocyte-secreted proteins that promote CNS synaptogenesis. *Cell* 120: 421–433.

Churchill, S. E. and F. H. Smith 2000a. A modern human humerus from the early Aurignacian of Vogelherdhöhle (Stetten, Germany). *American Journal of Physical Anthropology* 112: 251–273.

Churchill, S. E. and F. H. Smith 2000b. Makers of the early Aurignacian of Europe. *American Journal of Physical Anthropology* 113: 61–115.

Conard, N. J., P. M. Grootes and F. H. Smith 2004. Unexpectedly recent dates for human remains from Vogelherd. *Nature* 430: 198–201.

Cooper, A., C. Mourer-Chauvirp, G. K. Chambers, A. Von Haeseler, A. C. Wilson and S. Pääbo 1992. Independent origins of New Zealand moas and kiwis. *Proceedings of the National Academy of Sciences of the United States of America* 89: 8741–8744.

Cramer, H. 1941. Der Lebensraum des eiszeitlichen Höhlenbären und die 'Höhlenbärenjagdkultur'. *Zeitschrift der Deutschen Geologischen Gesellschaft* 93: 181–196.

Czarnetzki, A. 1983. Zur Entwicklung des Menschen in Südwestdeutschland. In H. Müller Beck (ed), *Urgeschichte in Baden-Württemberg*, pp. 217–240. Konrad Theiss, Stuttgart.

d'Errico, F. 1995. Comment on R. G. Bednarik, 'Concept-mediated markings of the Lower Palaeolithic'. *Current Anthropology* 36: 618–620.

Ehrenberg, K. 1951. 30 Jahre paläobiologischer Forschung in österreichischen Höhlen. *Quartär* 5: 93–108.

Ehrenberg, K. 1953a. Die paläontologische, prähistorische und paläoethnologische Bedeutung der Salzofenhöhle im Lichte der letzten Forschungen. *Quartär* 5: 35–40.

Ehrenberg, K. 1953b. Berichte über Ausgrabungen in der Salzofenhöhle im Toten Gebirge. VII. Beobachtungen und Funde der Salzofen-Expedition 1953. *Sitzungsberichte der Österreichischen Akademie der Wissenschaften, mathematisch-naturwissenschaftlicher Klasse* 162: 51–56.

Ehrenberg, K. 1954. Die paläontologische, prähistorische and paläo-ethnologische Bedeutung der Salzofenhöhle im Lichte der letzten Forschungen. *Quartär* 6: 19–58.

Ehrenberg, K. 1956. Berichte über Ausgrabungen in der Salzofenhöhle im Toten Gebirge. IX. Die Grabungen 1956 und ihre einstweiligen Ergebnisse. *Sitzungsberichte der Österreichischen Akademie der Wissenschaften, mathematisch-naturwissenschaftlicher Klasse* 165: 15–19.

Ehrenberg, K. 1957. Berichte über Ausgrabungen in der Salzofenhöhle im Toten Gebirge. VIII. Bemerkungen zu den Untersuchungen der Sedimente durch Elisabeth Schmid. *Sitzungsberichte der Österreichischen Akademie der Wissenschaften, mathematisch-naturwissenschaftlicher Klasse* 166: 57–63.

Ehrenberg, K. 1958. Vom dermaligen Forschungsstand in der Höhle am Salzofen. *Quartär* 10: 237–251.

Ehrenberg, K. 1959. Die urzeitlichen Fundstellen und Funde in der Salzofenhöhle, Steiermark. *Archaeologia Austriaca* 25: 8–24.

Ehrenberg, K. 1962. Über Lebensweise und Lebensraum des Höhlenbären. *Verhandlungen der Zoologisch-Botanischen Gesellschaft in Wien* 101: 18–31.

Ehrenberg, K. 1970. Vigaun, Salzburg. *Fundberichte aus Österreich* 9: 247.

Facorellis, Y., N. Kyparissi-Apostolika and Y. Maniatis 2001. The cave of Theopetra, Kalambaka: radiocarbon evidence for 50,000 years of human presence. *Radiocarbon* 43(2B): 1029–1048.

Fedele, F. G. and B. Giaccio 2007. Paleolithic cultural change in western Eurasia across the 40,000 BP timeline: continuities and environmental forcing. In P. Chenna Reddy (ed), *Exploring the mind of ancient man. Festschrift to Robert G. Bednarik*, pp. 292–316. Research India Press, New Delhi.

Fedele, F. G., B. Giaccio, R. Isaia and G. Orsi 2002. Ecosystem impact of the Campanian Ignimbrite eruption in Late Pleistocene Europe. *Quaternary Research* 57: 420–424.

Fedele, F. G., B. Giaccio, R. Isaia and G. Orsi 2003. The Campanian Ignimbrite Eruption, Heinrich Event 4, and Palaeolithic change in Europe: a high-resolution investigation. In *Volcanism and the earth's atmosphere*, pp. 301–325. Geophysical Monograph 139, American Geophysical Union, Washington, DC.

Felgenhauer, F. 1959. Das Paläolithikum von Willendorf in der Wachau, Niederösterreich. Vorbericht über die monographische Bearbeitung. *Forschungen und Fortschritte* 33(3): 152–155.

Foley, R. and M. M. Lahr 1997. Mode 3 technologies and the evolution of modern humans. *Cambridge Archaeological Journal* 7: 3–36.

Frayer, D. W. 1986. Cranial variation at Mladeč and the relationship between Mousterian and Upper Palaeolithic hominids. *Anthropos* (Brno) 23: 243–256.

Frayer, D. W. 1993. Evolution at the European edge: Neanderthal and Upper Paleolithic relationships. *Préhistoire Européenne* 2: 9–69.

Frayer, D. W. 1998. Perspectives on Neandertals as ancestors. In G. A. Clark and C. M. Willermet (eds), *Conceptual issues in modern human origins research*, pp. 220–234. Aldine de Gruyter, New York, NY.

Frayer, D. W., M. H. Wolpoff, F. H. Smith, A. G. Thorne and G. G. Pope 1993. The fossil evidence for modern human origins. *American Anthropology* 95: 14–50.

Frayer, D. W., M. H. Wolpoff, A. G. Thorne, F. H. Smith and G. G. Pope 1994. Getting it straight. *American Anthropology* 96: 424–438.

Fürst, J. 1922. Nové nálezy v Mladečých jeskyních. *Litovelské novinsky* 26 August.

Fürst, J. 1923-1924. Nálezy v jeskyních Mladečých. *Vlastivdný sborník střední a severní Moravy* 2: 6–7.

Gábori-Csánk, V. 1993. *Le Jankovichien: une civilisation paléolithiques en Hongrie*. ERAUL 53, Liège.

Gamble, C. 1999. *The Palaeolithic societies of Europe*. Cambridge University Press, Cambridge.

Gao, X. and C. J. Norton 2002. A critique of the Chinese 'Middle Palaeolithic'. *Antiquity* 76: 397–412.

Garrigan, D., Z. Mobasher, T. Severson, J. A. Wilder and M. F. Hammer 2005. Evidence for archaic Asian ancestry on the human X chromosome. *Molecular Biological Evolution* 22: 189–192.

Geay, P. 1957. Sur la découverte d'un squelette aurignacien en Charente-Maritime. *Bulletin de la Société Préhistoroque Française* 54: 193–197.

Geigl, E.-M. 2002. Why ancient DNA research needs taphonomy. Paper presented to Conférence ICAZ, 'Bioshere to Lithosphere', Durham.

Germonpré, M., M. V. Sablin, R. E. Stevens, R. E. M. Hedges, M. Hofreiter, M. Stiller and V. R. Després 2009. Fossil dogs and wolves from Palaeolithic sites in Belgium, the Ukraine and Russia: osteometry, ancient DNA and stable isotopes. *Journal of Archaeological Science* 36(2): 473–490.

Gibbons, A. 1998. Calibrating the mitochondrial clock. *Science* 279: 28–29.

Gibbons, A. 2010. Close encounters of the prehistoric kind. *Science* 328: 680–684.

Gieseler, W. 1974. *Die Fossilgeschichte des Menschen*. Konrad Theiss, Stuttgart.

Golenberg, E. M., A. Bickel and P. Weihs 1996. Effect of highly fragmented DNA on PCR. *Nucleic Acids Research* 24: 5026–5033.

González Echegaray, J., L. G. Freeman, I. Barandiaran, J. M. Apellaniz, K. Butzer, C. Fuentes Vidarte, B. Madariaga, J. A. Gonzalez Morales and A. Leroi-Gourhan 1980. *La Cueva de El Pendo*. Bibliotheca Praehistorica Hispana, Madrid.

Green, R. E., J. Krause, S. E. Ptak, A. W. Briggs, M. T. Ronan, J. F. Simons, L. Du, M. Egholm, J. M. Rothberg, M. Paunovic and S. Pääbo 2006. Analysis of one million base pairs of Neanderthal DNA. *Nature* 444: 330–336.

Green, R. E. et al. 2010. A draft sequence of the Neandertal genome. *Science* 328: 710–722.

Gutierrez, G. and A. Marin 1998. The most ancient DNA recovered from amber-preserved specimen may not be as ancient as it seems. *Molecular Biological Evolution* 15: 926–929.

Gutierrez, G., D. Sanchez and A. Marin 2002. A reanalysis of the ancient mitochondrial DNA sequences recovered from Neandertal bones. *Molecular Biological Evolution* 19: 1359–1366.

Gyllensten, U., D. Wharton, A. Josefsson and A. C. Wilson 1991. Paternal inheritance of mitochondrial DNA in mice. *Nature* 352: 255–257.

Hammer, M. F. 1995. A recent common ancestry for human Y chromosomes. *Nature* 378: 376–378.

Hardy, J., A. Pittman, A. Myers, K. Gwinn-Hardy, H. C. Fung, R. de Silva, M. Hutton and J. Duckworth 2005. Evidence suggesting that *Homo neanderthalensis* contributed the H2 *MAPT* haplotype to *Homo sapiens*. *Biochemical Society Transactions* 33: 582–585.

Harpending, H. C., M. A. Batzer, M. Gurven, L. B. Jorde, A. R. Rogers and S. T. Sherry 1998. Genetic traces of ancient demography. *Proceedings of the National Academy of Sciences of the United States of America* 95: 1961–1967.

Hasegawa, M., H. Kishino and T. Yano 1985. Dating of the human-ape splitting by a molecular clock of mitochondrial DNA. *Journal of Molecular Evolution* 22: 160–174.

Hawks, J. 1997. Have Neanderthals left us their genes? In L. Cavalli-Sforza (ed), *Human evolution: abstracts of papers presented at the 1997 Cold Spring Harbor Symposium on Human Evolution arranged by L. L. Cavalli-Sforza and J. D. Watson*, p. 81. Cold Spring Harbor Laboratory, Cold Spring Harbor, New York, NY.

Hawks, J., S.-H. Lee, K. Hunley and M. Wolpoff 2000. Population bottlenecks and Pleistocene human evolution. *Molecular Biological Evolution* 17: 2–22.

Hedges, S. B., S. Kumar, K. Tamura and M. Stoneking 1992. Human origins and analysis of mitochondrial DNA sequences. *Science* 255: 737–739.

Henke, W. and R. Protsch 1978. Die Paderborner Calvaria—ein diluvialer *Homo sapiens*. *Anthropologischer Anzeiger* 36: 85–108.

Henke, W. and H. Rothe 1994. *Paläoanthropologie*. Wiley-VCH Verlag GmbH, Berlin.

Henry-Gambier, D. 2002. Les fossiles de Cro-Magnon (Les-Eyzies-de-Tayac, Dordogne): Nouvelles données sur leur position chronologique et leur attribution culturelle. *Bulletin et Mémoires de la Société d'Anthropologie de Paris* 14(1–2): 89–112.

Higuchi, R., B. Bowman, M. Freiberger, O. A. Ryder and A. C. Wilson 1984. DNA sequences from the quagga, and extinct member of the horse family. *Nature* 312: 282–284.

Hublin, J.-J., C. B. Ruiz, P. M. Lara, M. Fontugne and J.-L. Reyss 1995. The Mousterian site of Zafarraya (Andalusia, Spain): dating and implications on the Palaeolithic peopling processes of western Europe. *Comptes Rendus de l'Academie des Sciences de Paris*, Series 2a 321(10): 931–937.

Hublin, J.-J., F. Spoor, M. Braun, F. Zonneveld and S. Condemi 1996. A late Neanderthal associated with Upper Palaeolithic artefacts. *Nature* 381: 224–226.

Jablonka, E. and M. Lamb 2005. *Evolution in four dimensions: genetic, epigenetic, behavioral, and symbolic variation in the history of life*. MIT Press, Cambridge, MA.

Jelínek, J. 1987. Historie, identifikace a význam mladečských antropologických nálezů z počátku mladého paleolitu. *Anthropos* 25: 51–69.

Jelínek, J., M. H. Wolpoff and D. W. Frayer 2005. Evolutionary significance of the Quarry Cave specimens from Mladeč. *Anthropologie* 43: 215–228.

Jolivet, J. P. and M. Henry 1994. *De la solution à l'oxyde, condensation des cations en solution aqueuse, chimie de surface des oxides*. InterEditions CNRS, Paris.

Karavanic, I. and H. Smith 1998. The Middle/Upper Palaeolithic interface and the relationship of Neanderthals and early modern humans in the Hrvatsko Zagorje, Croatia. *Journal of Human Evolution* 34: 223–248.

Kidd, K. K., J. R. Kidd, S. A. Pakstis, C. M. Tishkoff, C. M. Castiglione and G. Strugo 1996. Use of linkage disequilibrium to infer population histories. *American Journal of Physical Anthropology* Supplement 22: 138.

Klaatsch, H. and O. Hauser 1910. *Homo Aurignaciensis Hauseri*. *Prähistorische Zeitschrift* 1: 273–338.

Knies, J. 1906. Nový nález diluviálního človka u Mladče na Morav. *Vstník klubu přírodovdného v Prostjov* 7: 1–19.

Kozłowski, J. K. 1990. A multiaspectual approach to the origins of the Upper Palaeolithic in Europe. In P. Mellars (ed), *The emergence of modern humans. An archaeological perspective*, pp. 419–438. Edinburgh University Press, Edinburgh.

Krings, M., A. Stone, R. W. Schmitz, H. Krainitzki, M. Stoneking and S. Pääbo 1997. Neandertal DNA sequences and the origin of modern humans. *Cell* 90: 19–30.

Kuhn, S. L. 1995. *Mousterian lithic technology. An ecological perspective.* Princeton University Press, Princeton, NJ.

Kuhn, S. L. and A. Bietti 2000. The late Middle and early Upper Paleolithic in Italy. In O. Bar-Yosef and D. Pilbeam (eds), *The geography of Neandertals and modern humans in Europe and the Greater Mediterranean*, pp. 49–76. Peabody Museum of Archaeology and Ethnology, Cambridge, MA.

Kuhn, S. L. and M. C. Stiner 2001. The antiquity of hunter-gatherers. In C. Panter-Brick, R. H. Layton and P. Rowley-Conwy (eds), *Hunter-gatherers: an interdisciplinary perspective*, pp. 99–142. Cambridge University Press, Cambridge.

Kurtén, B. 1968. *Pleistocene mammals of Europe.* Weidenfeld and Nicolson, London.

Kyparissi-Apostolika, N. (ed) 2000. *Theopetra Cave. Twelve years of excavation and research 1987–1998.* Institute for Aegean Prehistory, Athens.

Kyrle, G. 1931. Die Höhlenbärenjägerstation. In O. Abel and G. Kyrle (eds), *Die Drachenhöhle bei Mixnitz*, pp. 804–962. Speläologische Monographien, Band 7–9, Vienna.

Lewin, R. 2005. *Human evolution: An illustrated introduction.* Blackwell, Oxford.

Lindhal, T. and B. Nyberg 1972. Rate depurination of native deoxyribonucleic acid. *Bio-chemistry* 11: 3610–3618.

Maddison, D. R., M. Ruvolo and D. L. Swofford 1992. Geographic origins of human mitochondrial DNA: phylogenetic evidence from control region sequences. *Systematic Biology* 41: 111–124.

Malez, M. 1956. Geoloska i paleontolska istrazivanja u pecini Veternici. *Acta Geologica Zagreb* 1: 83–88.

Malez, M. 1959. Das Paläolithikum der Veternicahöhle und der Bärenkult. *Quartär* 11: 171–188.

Malez, M. 1965. Novi opci varijacioni raspon vrste Ursus spelaeus Rosnm. et Heinroth. *Geoloski Vjesnik* 18: 133–139.

McDougall, I., F. H. Brown and J. G. Fleagle 2005. Stratigraphic placement and age of modern humans from Kibish, Ethiopia. *Nature* 433: 733–736.

Mellars, P. 2005. The impossible coincidence. A single-species model for the origins of modern human behavior in Europe. *Evolutionary Anthropology* 14: 12–27.

Mellars, P. and C. Stringer 1989. Introduction. In P. Mellars and C. Stringer (eds), *The human revolution: behavioural and biological perspectives on the origins of modern humans*, pp. 1–14. Edinburgh University Press, Edinburgh.

Morris, A. A. M. and R. N. Lightowlers 2000. Can paternal mtDNA be inherited? *The Lancet* 355: 1290–1291.

Mottl, M. 1950. Die paläolithischen Funde aus der Salzofenhöhle im Toten Gebirge. *Archaeologia Austriaca* 5: 24–34.

Movius, H. L. 1969. The Abri of Cro-Magnon, Les Eyzies (Dordogne) and the probable age of the contained burials on the basis of the nearby Abri Pataud. *Anuario de Estudio Atlanticos* 15: 323–344.

Napierala, H. and H.-P. Uerpmann 2010. A 'new' Palaeolithic dog from central Europe. *International Journal of Osteoarchaeology.* DOI: 10.1002/oa.1182

Nei, M. 1987. *Molecular evolutionary genetics.* Columbia University Press, New York, NY.

Noonan, J. P., M. Hofreiter, D. Smith, J. R. Priest, N. Rohland, G. Rabeder, J. Krause, J. C. Detter, S. Pääbo and E. M. Rubin 2005. Genomic sequencing of Pleistocene cave bears. *Science* 309: 597–599.

Oliva, M. 1989. Mladopaleolitické nálezy z Mladečských jeskyní. *Acta Musei Moraviae* 74: 35–54.

Orlando, L. and C. Hänni 2008. Du nouveau pour l'AND ancien. *Médicine Sciences* 8–9: I–XVI.

Pääbo, S. 1985. Molecular cloning and ancient Egyptian mummy DNA. *Nature* 314: 644–645.

Pääbo, S. 1989. Extraction, characterization, molecular cloning, and enzymatic amplification. *Proceedings of the National Academy of Sciences of the United States of America* 86: 1939–1943.

Pääbo, S., H. Poinar, D. Serra, V. Jaenicke-Després, J. Hebler, N. Rohland, M. Kuch, J. Krause, L. Vigilant and M. Hofreiter 2004. Genetic analyses from ancient DNA. *Annual Review on Genetics* 38: 645–679.

Palma di Cesnola, A. 1976. Le leptolithique archaïque en Italie. In B. Klíma (ed), *Périgordien et Gravettien en Europe*, pp. 66–99. Congrès IX, Colloque XV, UISPP, Nice.
Palma di Cesnola, A. 1989. L'Uluzzien: faciès italien du Leptolithique archaïque. *L'Anthropologie* 93: 783–811.
Pavlov, P., J. I. Svendsen and S. Indrelid 2001. Human presence in the European Arctic nearly 40,000 years ago. *Nature* 413: 64–67.
Pennisi, E. 1999. Genetic study shakes up out of Africa theory. *Science* 283: 1828.
Perpère, M. 1971. L'aurignacien en Poitou-Charentes (étude des collections d'industries lithiques). Doctoral thesis, University of Paris.
Perpère, M. 1973. Les grands gisements aurignaciens du Poitou. *L'Anthropologie* 77: 683–716.
Protsch, R. 1975. The absolute dating of Upper Pleistocene sub-Saharan fossil hominids and their place in human evolution. *Journal of Human Evolution* 4: 297–322.
Protsch, R. and H. Glowatzki 1974. Das absolute Alter des paläolithischen Skeletts aus der Mittleren Klause bei Neuessing, Kreis Kelheim, Bayern. *Anthropologischer Anzeiger* 34: 140–144.
Protsch, R. and A. Semmel 1978. Zur Chronologie des Kelsterbach-Hominiden. *Eiszeitalter und Gegenwart* 28: 200–210.
Pruvost, M., T. Grange and E.-M. Geigl 2005. Minimizing DNA contamination by using UNG-coupled quantitative real time PCR on degraded DNA samples: application to DNA studies. *Bio Techniques* 38(4): 569–575.
Pruvost, M., R. Schwarz, V. Bessa Correia, S. Champlot, S. Braguier, N. Morel, Y. Fernandez-Jalvo, T. Grange and E.-M. Geigl 2007. Freshly excavated fossil bones are best for amplification of ancient DNA. *Proceedings of the National Academy of Sciences of the United States of America* 104(3): 739–744.
Rabeder, G., D. Nagel and M. Pacher 2000. *Der Höhlenbär*. Species 4, Jan Thorbecke Verlag, Stuttgart.
Rakovec, I. 1967. Jamski medved iz Mokriske Jame v Savinjskih Alpah. *Razprave* 10: 123–203.
Ramirez Rozzi, F. V., F. d'Errico, M. Vanhaeren, P. M. Grootes, B. Kerautret and V. Dujardin 2009. Cutmarked human remains bearing Neandertal features and modern human remains associated with the Aurignacian at Les Rois. *Journal of Anthropological Sciences* 87: 153–185.
Relethford, J. H. 2002. Absence of regional affinities of Neandertal DNA with living humans does not reject multiregional evolution. *American Journal of Physical Anthropology* 115(1): 95–98.
Riel-Salvatore, J. and G. A. Clark 2001. Grave markers. Middle and Early Upper Paleolithic burials and the use of chronotypology in contemporary Paleolithic research. *Current Anthropology* 42: 449–479.
Rodriguez-Trelles, F., R. Tarrio and F. J. Ayala 2001. Erratic overdispersion of three molecular clocks: GPDH, SOD, and XDH. *Proceedings of the National Academy of Sciences of the United States of America* 98: 11405–11410.
Rodriguez-Trelles, F., R. Tarrio and F. J. Ayala 2002. A methodological bias toward overestimation of molecular evolutionary time scales. *Proceedings of the National Academy of Sciences of the United States of America* 99: 8112–8115.
Rougier H., Ş. Milota, R. Rodrigo, M. Gherase, L. Sarcină, O. Moldovan, R. G. Constantin, C. Franciscus, P. E. Zollikofer, M. Ponce de León and E. Trinkaus 2007. Peştera cu Oase 2 and the cranial morphology of early modern Europeans. *Proceedings of the National Academy of Sciences of the United States of America* 104(4): 1165–1170.
Schulz, H.-P. 2002. The lithic industry from layers IV–V, Susiluola Cave, western Finland, dated to the Eemian interglacial. *Préhistoire Européenne* 16–17: 7–23.
Schulz, H.-P., B. Eriksson, H. Hirvas, P. Huhta, H. Jungner, P. Purhonen, P. Ukkonen and T. Rankama 2002. Excavations at Susiluola Cave. *Suomen Museo* 2002: 5–45.
Schwartz, M. and J. Vissing 2002. Paternal inheritance of mitochondrial DNA. *New England Journal of Medicine* 347: 576–580.
Shang, H., H. Tong, S. Zhang, F. Chen and E. Trinkaus 2007. An early modern human from Tianyuan Cave, Zhoukoudian, China. *Proceedings of the National Academy of Sciences of the United States of America* 104(16): 6573–6578.

Skutil, J. 1938. *Pravěké nálezy v Mladči u Litovle na Morav*. Krajinská musejni společnost, Litovel.

Smith, F. H. 1982. Upper Pleistocene hominid evolution in south-central Europe: a review of the evidence and analysis of trends. *Current Anthropology* 23: 667–686.

Smith, F. H. 1985. Continuity and change in the origin of modern *Homo sapiens*. *Zeitschrift für Morphologie und Anthropologie* 75: 197–222.

Smith, F. H., I. Janković and I. Karavanić 2005. The assimilation model, modern human origins in Europe, and the extinction of Neandertals. *Quaternary International* 137: 7–19.

Smith, F. H. and G. Ranyard 1980. Evolution of the supraorbital region in Upper Pleistocene fossil hominids from south-central Europe. *American Journal of Physical Anthropology* 53: 589–610.

Smith, F. H., E. Trinkaus, P. B. Pettitt, I. Karavanić and M. Paunović 1999. Direct radiocarbon dates for Vindija G_1 and Velika Pećina Late Pleistocene hominid remains. *Proceedings of the National Academy of Sciences of the United States of America* 96(22): 12281–12286.

Smyčka, J. 1907. Litovel a okolí za pravku. *Pravk* 3: 140–150.

Smyčka, J. 1922. Nálezy diluviálního člověka v Mladči u Litovle na Moravě. *Obzor prehistorický* 1: 111–120.

Smyčka, J. 1925. Kdy přišel první člověk do litovelského kraje. Vlastivědný sborník střední a severní Moravy III, příloha Olomoucko a Litovelsko 6–7.

Soficaru, A., A. Doboş and E. Trinkaus 2006. Early modern humans from the Peştera Muierii, Baia de Fier, Romania. *Proceedings of the National Academy of Sciences of the United States of America* 103(46): 17196–171201.

Sonneville-Bordes, D. de 1959. Position-stratigraphique et chronologique relative des restes humains du Palaéolithique supérieur entre Loire et Pyrénées. *Annales de Paléontologie* 45: 19–51.

Stehlin, H. G. and A. Dubois 1916. *Note préliminaire sur les fouilles entreprises dans la Grotte de Cotencher (canton Neuchâtel)*. Ecologae Geologica Helvetia 14, Lausanne.

Stiner, M. C. 1994. *Honor among thieves. A zooarchaeological study of Neandertal ecology*. Princeton University Press, Princeton, NJ.

Strauss, E. 1999. Can mitochondrial clocks keep time? *Science* 283: 1435–1438.

Stringer, C. B. 1984a. Human evolution and biological adaptation in the Pleistocene. In R. Foley (ed), *Hominid evolution and community ecology: prehistoric human adaptation in biological perspective*, pp. 55–83. Academic, London.

Stringer, C. B. 1984b. The fate of the Neanderthals. *Natural History* (December): 6–12.

Stringer, C. B. 1985. Middle Pleistocene hominid variability and the origin of Late Pleistocene humans. In E. Delson (ed), *Ancestors: the hard evidence*, pp. 289–295. Alan R. Liss, New York, NY.

Stringer, C. B. 1989. The origin of early modern humans: a comparison of the European and non-European evidence. In P. Mellars and C. Stringer (eds), *The human revolution: behavioural and biological perspectives on the origins of modern humans*, pp. 232–244. Edinburgh University Press, Edinburgh.

Stringer, C. B. and P. Andrews 1988. Genetic and fossil evidence for the origin of modern humans. *Science* 239: 1263–1268.

Svoboda, J. 1990. The Bohunician. In J. K. Kozłowski (ed), *La mutation*, pp. 169–192. ERAUL, Liège.

Svoboda, J. 1993. The complex origin of the Upper Paleolithic in the Czech and Slovak Republics. In H. Knecht, A. Pike-Tay and R. White (eds), *Before Lascaux: the complete record of the early Upper Paleolithic*, pp. 23–36. CRC Press, Boca Raton, FL.

Svoboda, J. A., J. van der Plicht and V. Kuzelka 2002. Upper Palaeolithic and Mesolithic human fossils from Moravia and Bohemia (Czech Republic): some new ^{14}C dates. *Antiquity* 76: 957–962.

Szombathy, J. 1925. Die diluvialen Menschenreste aus der Fürst-Johanns-Höhle bei Lautsch in Mähren. *Die Eiszeit* 2: 1–34, 73–95.

Templeton, A. R. 1992. Human origins and analysis of mitochondrial DNA sequences. *Science* 255: 737.

Templeton, A. R. 1993. The 'Eve' hypothesis: a genetic critique and re-analysis. *American Anthropologist* 95: 51–72.

Templeton, A. R. 1996. Gene lineages and human evolution. *Science* 272: 1363.
Templeton, A. 2002. Out of Africa again and again. *Nature* 416: 45–51.
Templeton, A. R. 2005. Haplotype trees and modern human origins. *Yearbook of Physical Anthropology* 48: 33–59.
Terberger, T. 1998. Endmesolithische Funde von Drigge, Lkr. Rügen—Kannibalen auf Rügen? *Jahrbuch für Bodendenkmalpflege Mecklenburg-Vorpommern* 46: 7–44.
Terberger, T. and M. Street 2003. Jungpaläolithische Menschenreste im westlichen Mitteleuropa und ihr Kontext. In J. M. Burdukiewicz, L. Fiedler, W.-D. Heinrich, A. Justus and E. Brühl (eds), *Erkenntnisjäger: Kultur und Umwelt des frühen Menschen*, pp. 579–591. Veröffentlichungen des Landesamtes für Archäologie Sachsen-Anhalt – Landesmuseum für Vorgeschichte, Vol. 57/2, Halle.
Tobias, P. V. 1995. The bearing of fossils and mitochondrial DNA on the evolution of modern humans, with a critique of the 'mitochondrial Eve' hypothesis. *South African Archaeological Bulletin* 50: 155–167.
Torroni, A., M. T. Lott, M. F. Cabell, Y.-S. Chen, L. Lavergne and D. C. Wallace 1994. MtDNA and the origin of Caucasians: identification of ancient Caucasian-specific haplogroups, one of which is prone to a recurrent somatic duplication in the D-loop region. *American Journal of Human Genetics* 55: 760–776.
Trimmel, H. 1950. Die Salzofenhöhle im Toten Gebirge. Ph.D. thesis, Philosophical Faculty, University of Vienna.
Trinkaus, E. 2007. European early modern humans and the fate of the Neandertals. *Proceedings of the National Academy of Sciences of the United States of America* 104: 7367–7372.
Trinkaus, E. and M. Le May 1982. Occipital bunning among Later Pleistocene hominids. *American Journal of Physical Anthropology* 57: 27–35.
Trinkaus, E., O. Moldovan, Ş. Milota, A. Bîlgar, L. Sarcina, S. Athreya, S. E. Bailey, R. Rodrigo, G. Mircea, T. Higham, C. Bronk Ramsey and J. van der Plicht 2003. An early modern human from the Peştera cu Oase, Romania. *Proceedings of the National Academy of Sciences of the United States of America* 100(20): 11231–11236.
Tschumi, O. 1949. Die steinzeitlichen Epochen. In O. Tschumi (ed), *Urgeschichte der Schweiz*, Vol. 1, pp. 407–727. Verlag Huber und Co., Frauenfeld.
Vaquero, M., M. Esteban, E. Allué, J. Vallverdú, E. Carbonell and J. L. Bischoff 2002. Middle Palaeolithic refugium, or archaeological misconception? A new U-series and radiocarbon chronology of Abric Agut (Capellades, Spain). *Journal of Archaeological Science* 29: 953–958.
Vértes, L. 1951. Novi'e raskopki v peschtschere na Istállóskö. *Acta Archaeologica* 1: 15–34.
Vértes, L. 1955. Neuere Ausgrabungen und paläolithische Funde in der Höhle von Istállóskö. *Acta Archaeologica* 5: 111–131.
Vértes, L. 1959. Die Rolle des Höhlenbären im ungarischen Paläolithikum. *Quartär* 11: 151–170.
Vértes, L. 1965. *Az Öskökor és az tmeneti kökor emlékei Magyarorsz gon.* Akadémiai Kiadó, Budapest.
Vigilant, L., M. Stoneking, H. Harpending, K. Hawkes and A. C. Wilson 1991. African populations and the evolution of human mitochondrial DNA. *Science* 253: 1503–1507.
Wainscoat, J. 1987. Out of the garden of Eden. *Nature* 325: 13.
Wainscoat, J. S., A. V. S. Hill, A. L. Boyce, J. Flint, M. Hernandez, S. L. Thein, J. M. Old, J. R. Lynch, A. G. Falusi, D. J. Weatherall and J. B. Vlegg 1986. Evolutionary relationships of human populations from an analysis of nuclear DNA polymorphisms. *Nature* 319: 491–493.
Walberg, M. W. and D. A. Clayton 1981. Sequence and properties of the human KB cell and mouse L cell D-loop regions of mitochondrial DNA. *Nucleic Acids Research* 9: 5411–5421.
Walser, J. and A. Furano 2010. The mutational spectrum of non-CpG DNA varies with CpG content. *Genome Research.* DOI: 10.1101/gr.103283.109
Watchman, A. 1999. A universal standard for reporting the ages of petroglyphs and rock paintings. In M. Strecker and P. Bahn (eds), *Dating and the earliest known rock art*, pp. 1–3. Oxbow Books, Oxford.
Weidenreich, F. 1946. *Apes, giants, and man.* University of Chicago Press, Chicago.

Weiser, E. 1928. *Reise und Wanderbuch*, 2nd volume. Ernst Feld, Freudenthal.

White, R. 1993. Technological and social dimensions of Aurignacian-age body ornaments across Europe. In H. Knecht, A. Pike-Tay and R. White (eds), *Before Lascaux: the complex record of the early Upper Palaeolithic*, pp. 277–299. CRC Press, Boca Raton, FL.

White, R. 1995. Comment on R. G. Bednarik, 'Concept-mediated markings in the Lower Palaeolithic'. *Current Anthropology* 36: 623–625.

White, T. D., B. Asfaw, D. DeGusta, H. Gilbert, G. D. Richards, G. Suwa and F. C. Howell 2003. Pleistocene *Homo sapiens* from Middle Awash, Ethiopia. *Nature* 425: 742–747.

Wild, E. M., M. Teschler-Nicola, W. Kutschera, P. Steier, E. Trinkaus and W. Wanek 2005. Direct dating of Early Upper Palaeolithic human remains from Mladeč. *Nature* 435: 332–335.

Williams, R. S. 2002. Another surprise from the mitochondrial genome. *New England Journal of Medicine* 347: 609–611.

Wolpoff, M. 1999. *Paleoanthropology*, 2nd edn. McGraw-Hill, New York, NY.

Wolpoff, M., F. H. Smith, M. Malez, J. Radovčić and D. Rukavina 1981. Upper Pleistocene hominid remains from Vindija Cave, Croatia, Yugoslavia. *American Journal of Physical Anthropology* 54: 499–545.

Zischler, H., H. Geisert, A. von Haeseler and S. Pääbo 1995. A nuclear 'fossil' of the mitochondrial D-loop and the origin of modern humans. *Nature* 378: 489–492.

Zotz, L. F. 1939. *Die Altsteinzeit in Niederschlesien*. Kabitsch Verlag, Leipzig.

Zotz, L. F. 1944. *Altsteinzeitkunde der Südostalpenländer*. Archiv für vaterländische Geschichte und Topographie 29, Weimar.

Zotz, L. F. 1951. *Altsteinzeitkunde Mitteleuropas*. F. Enke, Stuttgart.

Chapter 3
The Hard Evidence

The aim of science is not to open the door to infinite wisdom,
but to set a limit to infinite error
(Brecht 1980: sc. 9).

Investigating Iconicity

Any thoughtful person will, upon reflection, arrive at the opinion that humans became human not through natural processes that modified their skeletal structures, but by processes that enabled them to develop culture, cognition, and technology on a scale significantly separating humans from all other primates in those areas. Pleistocene archeologists, on the other hand, have instead developed models that define stone tools or their assemblages as "culture." Apart from the inherent inability to test this taxonomy, tools do not define cultures, as we have noted. Therefore, the "cultural sequence" archeology has given us of the Pleistocene should not be expected to be a sequence of real cultures. Similarly, the history of the somatic evolution of hominins is fascinating, but it does not adequately define the process of humanization, to which changes in skeletal details are only peripheral. Yet the efforts to learn about the cognitive, neurological, and cultural evolution of hominins have so far not been of adequate depth, and the processes involved in it have not been clarified.

Archeological narratives have thus become biased in favor of interpreting those variables the discipline perceived itself best equipped to deal with. As a result of this, and of archeological reluctance to adopt taphonomic logic (Bednarik 1994a), the dominant narratives of the Paleolithic periods of human history—the periods we are concerned with here—are more far-fetched and probably more invalid than those of any other period of our existence as a genus. The dominant dogma developed for these periods over the past century perceives little or no cultural change or evolution throughout the Lower Paleolithic. It defines culture of this time period as static, and sees little change even in the subsequent Mode 3 technocomplex ("Middle Paleolithic"; Foley and Lahr 1997). Then, with the advent of the Upper Paleolithic, the dogma perceives the cataclysmic "bottleneck," the "quantum jump'

R.G. Bednarik, *The Human Condition*, Developments in Primatology: Progress and Prospects, DOI 10.1007/978-1-4419-9353-3_3,

or "explosion" we have visited in Chapter 2. Accordingly, all the typically human characteristics that distinguish us from other animals appeared suddenly and at once—and, of course, first in western Europe: art, language, complex social systems, self-awareness, forward planning, and symboling. This seems to repeat the state in paleoanthropology early in the twentieth century, when the Piltdown fake was readily accepted as demonstrating that humans originate in England. In reality, Europe is an insignificant appendage of Asia, on the margins of the main theatres of human evolution, and for that reason alone the notion that characteristics of human modernity evolved there first should be highly suspect. Whereas in paleoanthropology the Eurocentrism has been corrected since the middle of the twentieth century, this correction is still to occur in cultural history.

The dominant model is rejected here on the basis of the hard evidence, according to which the development of human cognition, like that of physical evolution and encephalization, was a gradual process that occurred throughout the Pleistocene period. This evidence consists of indications of complex premeditated human behavior (such as the colonization of land only accessible by sea crossings) and the cognition (e.g., language ability) and technology demanded by it; the use of coloring materials and manuports; the use and making of beads and pendants; the production of petroglyphs and portable engravings; and the introduction of the concept of iconicity (of referrer and referent). Since the last-named is perhaps the one most directly related to defining the origins of symboling we begin with this crucial generic factor.

Iconicity is the property of a marking or shape that provides visual information recognized by most contemporary humans as resembling the form of an object. A marking or object (referrer) is considered iconic when most modern people tend to see it as resembling a different object (referent). The susceptibility of modern humans to pareidolia, i.e., to perceiving visual similarities between non-iconic entities (rocks, clouds, or whatever) and their referents varies very greatly among individuals. However, iconic resemblance of a referent is not self-evident; its detection requires an appropriate perceptual mechanism. Visual ambiguity, from which this facility probably developed (Bednarik 2003a), is a property widely experienced by species throughout the animal kingdom, but it is thought that only hominins developed a cultural use of this feature. Examples from the animal world include eye-like markings on the wings of moths and numerous forms of camouflage, or the reactions of countless animals to the likeness of a raptor or other carnivore. The experience of perceiving, for an instant, a snake on a forest path when in fact there is only an exposed tree root is an example of visual ambiguity (cf. Coss 1985: 256; Pinker 1997: 386; Watson 2009: 163), which seems to prompt an alert-reaction deriving from an archaic, hard-wired neuronal template (Bednarik 1986). Such visual misidentification (Bednarik 2003a) might in an organism capable of "conscious" reflection lead to perceiving a connection between referent and referrer (or the signified and the signifier). In this theory, the actual production of iconographic forms becomes the cultural and intentional creation of features prompting visual responses to a signifier; *it induces visual ambiguity intentionally*. The lines, colors, and textures making up an image adequately resemble abstracted aspects of the object portrayed to "persuade" the visual system's recall to "see the resemblance." This definition of art may sound overly esoteric but it is perhaps crucial in effectively understanding the nature and origins of iconographic or figurative art; it is also crucial in understanding hominin cognition and symboling.

In iconic symbolism, the connection between referent and referrer is via iconicity. This is a relatively simple form of symboling, in the sense that an organism capable of cognitively perceiving visual ambiguity detects at least some meaning without any cultural (i.e., learned) faculties coming into play. The cognition involved is deeply rooted in mental processes found in numerous animal species, such as flight reactions to the silhouette of a bird of prey, i.e., facilities encoded in DNA. Rats retrieve suitably sized objects if they bear three dots resembling eyes and a nose and hide it, as if it were an infant rat; they seem to react automatically to the "face" pattern. The relevant cognitive processes are even related to the effect of camouflage, which is just as widespread in natural systems. Some animal species of sophisticated cognitive faculties master iconic recognition, in the sense that they recognize a likeness in a photograph or film (Cabe 1980: 324–325). This includes chimps, bonobos, and orangutans. The incredible mastery of this ability, as well as some other startling insights, are evident from the observation of one chimp that, upon finding in a magazine a photograph of a nude female human, began to masturbate. Indeed, if we are to learn about the way our own cognition operates, we would do well to pay particular attention to primate behavioral studies.

It is, therefore, essential to perceive symbolism based on iconicity as cognitively much more rudimentary than a symbolism requiring the link between referent and referrer to be negotiated culturally. The use of iconic symbols is not as effective in promoting cognitive flexibility as, for instance, linguistic symbols (Jacques and Zelazo 2005: 149), a point we will return to in Chapter 6. For instance, a bead is a noniconic object that has exceedingly complex symbolic roles; it stores a wealth of cognitive information outside the human brain (Gregory (1970: 148; Goody 1977; Donald 1991: 124–161), yet its meaning is only accessible to an organism possessing the "software" of the cultural conventions concerned: it takes a meaning to catch a meaning. The same, obviously, applies to any anthropic marking that is imbued with meaning not detectable iconographically, such as a geometric motif or other noniconic graphic symbol. The acoustic or phonetic equivalent of iconicity is onomatopoeia, which refers to the formation of words by imitating a sound associated with the referent. Typical onomatopoeic words are "cuckoo" or "buzz." With them the meaning is either obvious, or detecting it requires only minimal cultural (learned) cues.

In much the same way there are forms of modified iconicity: natural forms whose iconic qualities have been emphasized by anthropic modification. This observation leads to a fundamental differentiation between three forms of symbolism in paleoart: iconic, modified iconic, and noniconic (Bednarik 2003a; Sreenathan et al. 2008). The most direct is by iconicity of purely natural, i.e., unmodified forms. It occurs when an object of the natural world offers sufficient visual clues to prompt the mental bridge to be made between referent and referrer. In the earliest paleoart, typical representatives of this phenomenon are manuports such as the Makapansgat cobble

(Bednarik 1998) or the Erfoud Site fossil cast (Bednarik 2002a), which are of such effective iconic properties that they were noticed by hominins up to three million years ago. Such objects attracted sufficient curiosity to be collected and taken back to occupation sites, without being modified.

The ability of detecting such levels of iconicity is certainly not very much beyond the capability of chimps or bonobos, or possibly even within it, so it is reasonable to expect it in australopithecines and earliest hominins, such as *Kenyanthropus platyops* (3.5 Ma BP). The jasperite cobble from the dolomite cave Makapansgat (Fig. 3.1), near Potgietersrus, northern South Africa, was excavated by W. I. Eitzman in 1925 from the level 3 pink stony bone breccia (2.5–3 Ma old; McFadden et al. 1979), which also contained numerous australopithecine remains as well as a few stone tools (Eitzman 1958). Dart (1974) and Oakley (1981) considered it to be a manuport carried into the cave because of its striking visual properties: its red color and the distinctive markings. The stone's first detailed examination (Bednarik 1998) refuted the opinion that the markings may have been emphasized or even entirely made by australopithecines (Bahn 1997). The markings are natural features and a complete microscopic scan revealed no trace of intentional modification. Nevertheless, the cobble derives from an alluvial deposit and was certainly carried over a considerable (but unknown) distance before being deposited in the cave. Initially formed by energetic fluvial transport from heterogeneous jasperite, the cobble became embedded in a silicified conglomerate comprising mostly well-sorted quartz sand and silt-grade sediment. Upon weathering out of that facies it ended up again in a river and was marked by suspended particles (for a reconstruction

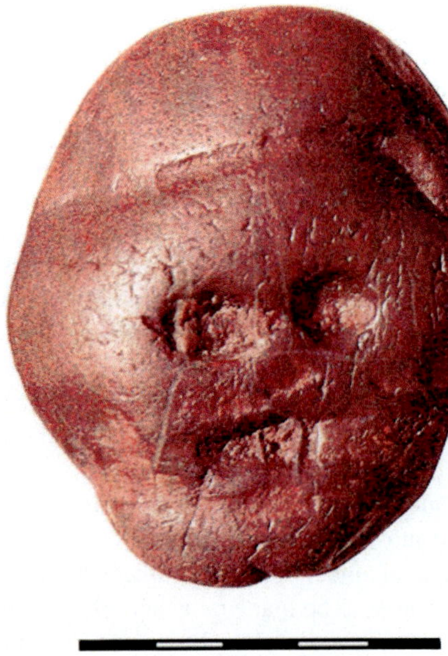

Fig. 3.1 Jasperite cobble from Makapansgat, South Africa, deposited in an australopithecine-bearing cave sediment almost three million years ago; scale in cm

of the specimen's history, see Bednarik 1998). The presence of stone tools in the cave's breccia raises the possibility that it was carried into the cave not by australopithecines, but by some of the earliest hominins.

Similarly, microscopic examination of the Erfoud Site A-84-2 cuttlefish fossil cast yielded no evidence of anthropic modification (Bednarik 2002a). Found in a dense cluster of Late Acheulian lithics within an arrangement of rocks thought to represent a dwelling, the Devonian or Carboniferous fossil (*Orthoceras* sp.) resembles a perfectly naturalistic and life-size, nonerect human penis. The site is located in the vicinity of the townships Erfoud and Rissani, eastern Morocco, and the type of fossil, although common in other parts of the country, does not occur naturally in the region. This 67.4-mm-long manuport of a light brown and semitranslucent chalcedonic silica has been carried for a very considerable distance, almost certainly because of its appearance. It bears the same mineral accretion as the stone tools found with it (Fig. 3.2).

Another early example of possible direct iconographic symbolism is via fossil casts, of both floral (e.g., ferns) and faunal specimens (Feliks 1998). Fossils, such as shells, are a prime example of a class of natural forms offering many, if not most, of the visual characteristics of the referent (the live organism, in this case). It seems possible that hominins benefited cognitively from making the connection between referrer and referent in such obvious cases, which might explain why they seem to have "curated" such objects (e.g., Oakley 1981). This could have prompted the establishment of neural pathways permitting the understanding that one thing can stand for another, as well as the appreciation that the objects of the object world can be grouped into classes on the basis of taxonomic criteria (Bednarik 1990a). These two abilities were among the most important cognitive milestones in human evolution; therefore, they need to be investigated here in some detail. The evidence suggests that both appeared in the Early Pleistocene. It is hardly a coincidence that their advent was followed by the quantum jump in technological capacities apparent in the late part of the Early Pleistocene—a development which orthodox Pleistocene archeology has yet to discover.

Fig. 3.2 The Erfoud manuport, from a Late Acheulian dwelling in the Moroccan Sahara; scale in mm

Paleoart of the Lower Paleolithic

The discipline's preoccupation with the "short range" proposition of a cognitive "explosion" around 40 ka ago has thrown a long shadow over the preceding part of the human journey. It has effectively prevented the acceptance of any form of advanced behavior or accomplishment by previous hominins, such as their incredible colonizations across the sea, or their use of complex symbolisms. Evidence of some of these feats extends back in the order of 800 ka to one million years ago; others appear first manifested during the course of the Middle Pleistocene period (780–127 ka ago).

Toward the end of the Early Pleistocene, hominins apparently began to discriminate between "exotic" articles (such as crystal prisms, fossil casts, unusually shaped, or colored stones) and "ordinary" ones (Bednarik 1990a; Dissanayake 1988). For instance, the collection of crystal prisms, often much too small to have served as a source for stone tool material, has been noted in several Mode 1 and Mode 2 assemblages: Zhoukoudian, China (with *Homo erectus* remains; Pei 1931: 120); Singi Talav, India (Lower Acheulian; d'Errico et al. 1989); Gesher Benot Ya'aqov, Israel (Goren-Inbar et al. 1991); Gudenushöhle, Austria (Bednarik 1988); and Wonderwerk Cave, South Africa, up to 800 ka ago (Bednarik 1993a). It was also in the earliest part of the Middle Pleistocene that hominins left the very first evidence of one of the most important indicators of symboling, the use of pigment (Bednarik 1990b, 1992, 1994b, 2003b). This may roughly coincide with the expansion of humans into Europe (agreement on the timing of this event is still elusive, however, and several pre-1 Ma dates have been proposed, including the earliest, of 1.57 Ma at Lézignan-la-Cèbe, southern France; Crochet et al. 2009), possibly via the Strait of Gibraltar (Bednarik 1999a); and certainly does so with the introduction of seafaring in Wallacea, Indonesia (Bednarik 1999b, 2001a, 2003c).

Pigment

Evidence of the very early use of iron oxides and hydroxides, presumably as coloring matter, has long been demonstrated from many sites in the Old World. Finds of hematite and similar minerals that bear striation use marks are known from several occupation sites of very early periods, in various parts of Africa, Europe, and India (Bednarik 1992, 1994b).

Some of the earliest relevant evidence is provided by Wonderwerk Cave in South Africa, where human occupation evidence begins before 1.8 Ma ago and numerous ochre fragments occur at many of its levels (Imbrie et al. 1984; Beaumont 1990, 1999; Binneman and Beaumont 1992; Bednarik 1994b). Older still are the two lumps of "ochre" Leakey (1958) has reported from the Developed Oldowan of Olduvai BK 2, Tanzania, but they were subsequently identified as red volcanic tuff (Oakley 1981: 207) and are questionable evidence. A hematite piece from Kabwe Cave near Broken Hill, Zambia, is probably in the order of 300 ka old, and there is a spheroid stone of 60 mm with red staining from the same site to be considered as well (Clark et al. 1947). Clark (1974) also reports evidence of pigment use from the Acheulian site at Kalambo Falls, Zambia, which is probably around 200 ka old. The same age applies to the red and yellow ochre lumps found with a grinding slab of the Lower Sangoan, from Sai Island, Sudan (Van Peer et al. 2003). Somewhat older than that is a ground piece of hematite from Nooitgedacht, South Africa (Beaumont and Morris 1990). The red pigment traces on the Tan-Tan figurine from Morocco also need to be considered in this context, even though they are only microscopic, but at around 400 ka they do represent the earliest evidence of *applied* pigment that we currently have (Bednarik 2001b, 2003a).

Whereas the more recent Middle Stone Age has long yielded major quantities of iron pigments in southern Africa, including quite extensive mining evidence (Stapleton and Hewitt 1928; Beaumont and Boshier 1972; Beaumont 1973; Miller et al. 1999; Grün and Beaumont 2001; Henshilwood et al. 2001, 2002), these earlier African finds remained relatively isolated. However, recently the quantity of such material available from the Lower Stone Age of sub-Saharan Africa has increased significantly, and with it the evidence of its use in the form of striation facets. This includes more than seventy red ochre pieces, over 5 kg in total weight, from site GnJh-15 in the Kapthurin Formation, Kenya, >285,000 years old (McBrearty 2001: 92). More than 306 pieces of specularite, hematite, limonite, ochrous sandstone, and manganese dioxide have been excavated at Twin Rivers, Zambia, dated to between 270 and 170 ka BP; 3% of this material shows signs of modification by grinding or rubbing (Barham 2002).

This confirms the actual use of ferruginous pigment during the Lower Paleolithic period, first demonstrated by Marshack (1981) in Europe and by me in Asia. Marshack has reported a 33-mm hematite piece from the Acheulian of Beçov, Czech Republic, striated on two faces. The floor near this find was covered by pigment powder, suggesting an activity of manufacturing coloring powder at this site. Among a series of almost twenty hematite pebbles found in the Lower Acheulian layer of Hunsgi, India, one 20-mm specimen bears a distinct facet with subparallel striations indicative of its use as a crayon to color a rock surface (Bednarik 1990b). We cannot know what these color markings may have looked like, but the evidence that they must have been made raises the possibility that there was some form of pigmented rock art. A few European Acheulian sites had earlier yielded tentative evidence of ochre use, including Terra Amata, France, where several apparently facetted fragments were noticed among 75 pieces of red, brown, and yellow, fire-treated limonite deposited about 380 ka ago (de Lumley 1966). A reportedly shaped slab of ochre was also found in the Acheulian of Ambrona, Spain (Howell 1966: 129), and a "rubbed" hematite fragment from Achenheim, France, seems to be about 250 ka old (Thévenin 1976).

These finds imply that pigments have been in use for much if not all of the Middle Pleistocene of southern Africa, and elsewhere in the Old World for at least much of the second half of that period. Ochre and similar minerals can be used for body painting, for the painting of objects (as indicated in the Tan-Tan proto-figurine) or to draw on surfaces, notably on rock. All of these activities demand complex cultural practices and probably the use of symbolism.

Petroglyphs

Painted designs are extremely unlikely to survive from these early times, in fact at unprotected sites they could not survive from the entire Pleistocene. Petroglyphs, however, have significantly better prospects, especially if they occur on the most weathering-resistant rock types, and preferably in sheltered locations. Petroglyphs relating to "Middle Paleolithic" traditions are very common; they are more numerous than "Upper Paleolithic" rock art motifs (Bednarik 1995a: 628). Nevertheless, the number of petroglyphs credibly attributed to the Lower Paleolithic period remains relatively small, but it must be remembered that nearly all examples refer to discoveries of recent years.

The first rock art ascribed to the Lower Paleolithic are the eleven petroglyphs in Auditorium Cave, Bhimbetka complex, Madhya Pradesh, India (Bednarik 1993b, 1994c). Nine cupules (hemispherical cup-marks, the most common form of petroglyph in the world; Bednarik 2008) occur on a large vertical boulder face above ground level (Fig. 3.3), while a tenth cupule and a meandering groove clearly associated with it were found in an excavation, covered by the uppermost part of substantial Late Acheulian occupation deposits (Fig. 3.4). The latter were overlain by a horizon of calcite-cemented "Middle Paleolithic" sediment that virtually excludes the possibility of postdepositional disturbance. Below the Acheulian deposits, and separated from them by a sterile layer, an Olduwan-like industry of chopping tools forms the basal feature. The cave is in heavily metamorphosed quartzite, a rock of such hardness that it was extensively quarried by hominins at several Bhimbetka sites.

Another Indian quartzite cave, Daraki-Chattan, was found to contain two vertical panels densely covered by 498 cupules (Kumar 1996). Because apparently Middle Paleolithic and Acheulian lithics occur on the surface of the cave's floor deposit,

Fig. 3.3 Chief's Rock, the central feature of Auditorium Cave, Bhimbetka site complex, central India. Some of the nine Paleolithic cupules on the vertical panel are visible

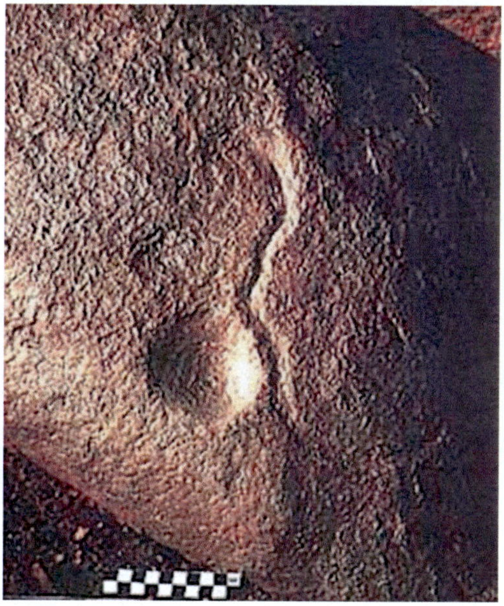

Fig. 3.4 Auditorium Cave at Bhimbhetka, India, excavated Lower Paleolithic cupule and meandering groove petroglyphs

Giriraj Kumar suggested that these cupules might be of great age as well (Fig. 3.5). In response to these and some other discoveries, the Early Indian Petroglyphs (EIP) Project was established, with the intention of testing these claims by an international panel of specialists (Bednarik 2000; Kumar et al. 2003). As part of the EIP Project, major excavations were commenced at Bhimbetka and Daraki-Chattan in 2002. At the latter site this led to the excavation of numerous exfoliated wall

Fig. 3.5 Lower Paleolithic cupules on the wall of Daraki-Chattan cave, central India

fragments found within the Pleistocene occupation deposit. These rock slabs bear a total of twenty-eight further cupules, identical to those on the walls above. They appeared throughout the sediment strata, almost down to bedrock, extending well into a deposit of chopping tools (Fig. 3.6). That layer also contained numerous hammerstones that had been used to create the cupules most laboriously. Replication experiments determined that in excess of 30,000 blows with the hand-held tools are required to create one such cupule on the rock in question; hence the effort invested in creating this rock art site, during a pre-Acheulian period, bears an incredible witness to the importance of such symboling activity to the people concerned.

While some Indian sites thus present the currently oldest known rock art in the world, there are also a few southern African finds that need to be mentioned here. First, there is the phonolite cobble Leakey (1971: 269, Pl. 17) reported from Floor FLK North 1 in Bed 1, Olduvai Gorge. The 10.5-cm specimen is artificially grooved and pecked, bearing what appears to be one cupule on each side. However, cupule-like features on portable rock have been produced by chimps and other primates, resulting from such activities as cracking nuts (McGrew 1992: 205, 1993), and Joulian (1995: figure 5) presents a *percuteur* made by chimps that resembles

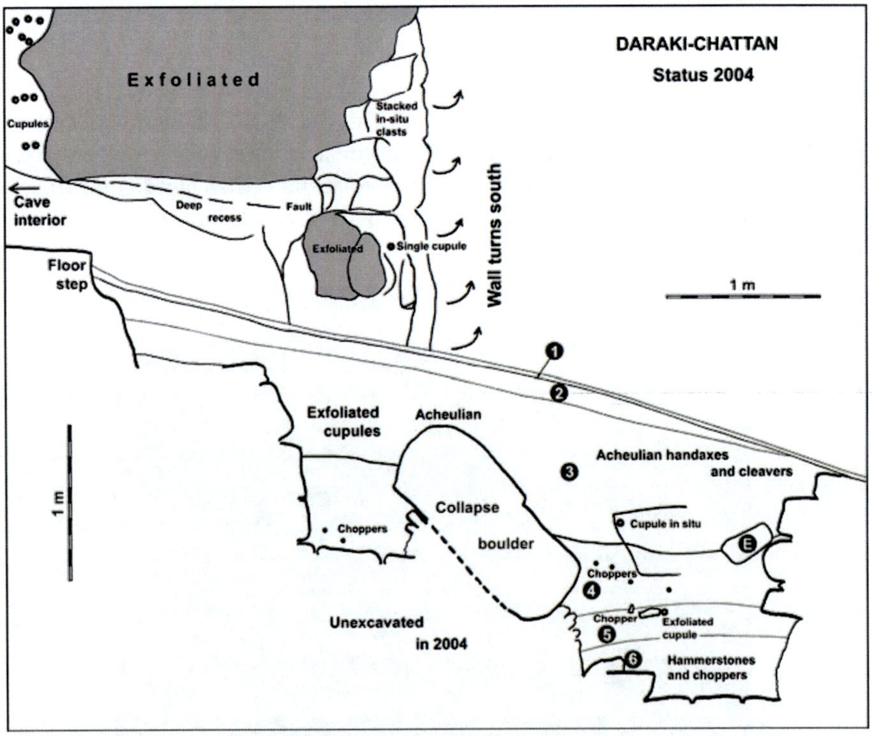

Fig. 3.6 Stratigraphy of Daraki-Chattan cave, showing the distribution of exfoliated rock slabs with cupules in the Mode 2 and Mode 1 sediments, and the hammerstones used in cupule production in the Mode 1 occupation deposits

Leakey's specimen. Bearded capuchin monkeys (*Cebus libidinosus*) at Boa Vista, Brazil, have produced hollows and groups of mortar-like, quite deep percussion pits on horizontal rock surfaces through their skilled use of rock mauls in cracking palm nuts. These panels can closely resemble groups of mortars or large cupules.

A second relevant find in South Africa is a grindstone of the Fauresmith industry bearing a partly pecked grid pattern. It was reported by Laidler (1933) from Blind River Mouth in East London. The Fauresmith, characterized by small well-made handaxes, is a Late Acheulian industry in the interior of southern Africa.

In 2001, Peter Beaumont discovered a series of very early cupule sites in the Korannaberg region of the southern Kalahari (Fig. 3.7). Like very early Indian cupules, these are also found on particularly hard quartzites. Nearby found artifacts are of the Middle Stone Age, Fauresmith, and Acheulian. An examination of the sites in 2009 suggested that most of the cupules as well as a good number of circle petroglyphs are of the Middle Stone Age, but at two of the localities, Nchwaneng and Potholes Hoek, these may be preceded by a small number of Fauresmith cupules (Beaumont and Bednarik 2010).

Finally, a Nubian sandstone slab has been excavated, c. 60 cm long, bearing a distinct grinding hollow or large cupule of about 10 cm, around which seven very small cupules (c. 1 cm diameter) are arranged. Found with red and yellow ochre lumps, this object is from Sai Island, Sudan, and belongs to the Lower Sangoan, being about 200 ka old (Van Peer et al. 2003) (Fig. 3.8).

Proto-figurines

No doubt these finds will encourage interest in locating more petroglyphs of the Lower Paleolithic, but for the time being they are the only examples reported.

Fig. 3.7 Cupules of the Fauresmith or Middle Stone Age tradition, on the 300-million-year-old glacial pavement of Potholes Hoek, Korannaberge, South Africa

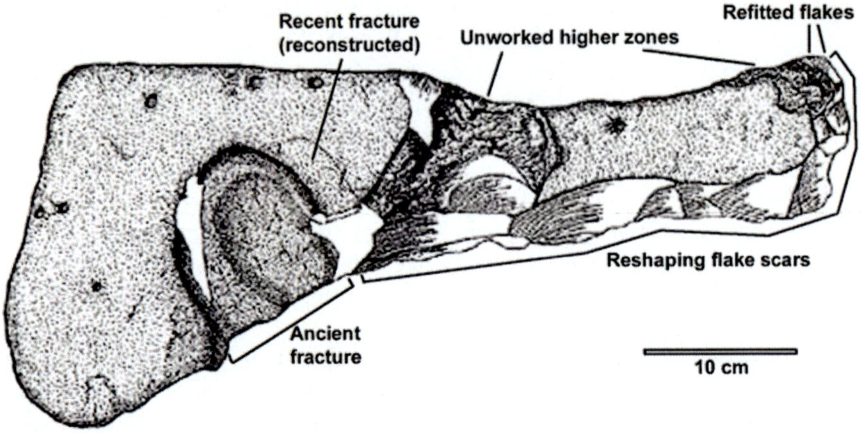

Fig. 3.8 Cupules from Sai Island, Sudan, Lower Sangoan, about 200 ka old (redrawn after Van Peer et al. 2003)

Similarly, the possibility of locating proto-figurines from that period has also only been considered in recent years, and currently only two specimens appear to deserve that designation. This requires evidence that they are not just iconic, in the sense that they resemble another object they are seen to represent; there must also be a clear indication that the object was modified by human hand so as to emphasize that natural or inherent iconicity.

A basaltic tuff pebble containing scoria clasts was excavated from a large occupation deposit of the Late Acheulian at Berekhat Ram, Israel, and is older than 230 ka (Goren-Inbar 1985). Its natural form, suggestive of the head, torso, and arms of a female human (Fig. 3.9), has been emphasized by man-made grooves implying that the iconic properties of the object were appreciated (Goren-Inbar 1986; Goren-Inbar and Peltz 1995). Most commenting authors rejected the find in the subsequent years without examining it (e.g., Chase and Dibble 1987; Davidson 1990; Pelcin 1994; Nowell 1995; Noble and Davidson 1996: 75; Davidson and Noble 1998). Marshack (1996, 1997) conducted a microscopic study of the object's markings, concluding that the grooves and abrasions were made with stone tools. His main findings were then corroborated by d'Errico and Nowell (2000). They accepted the object's artifact status, but they still queried the significance of its iconicity—even though they called it a "figurine." Another issue of continuing concern was the unique status of the Berekhat Ram object.

Both these issues were resolved shortly later with the report of a second stone figurine from the near-Mediterranean region. The object from Tan-Tan, Morocco, is of quartzite and comes from a Middle Acheulian occupation layer thought to be about 400 ka old on the basis of the accompanying lithic typology (Bednarik 2001b, 2003a). Its anthropomorphous form is much more pronounced than that of the Israeli specimen, with both arms and legs well represented, and it is emphasized by eight symmetrically arranged grooves (Fig. 3.10). Five of these lines were found to have

Fig. 3.9 Late Acheulian proto-figurine, Berekhat Ram, Israel, probably in the order of 300,000 years old

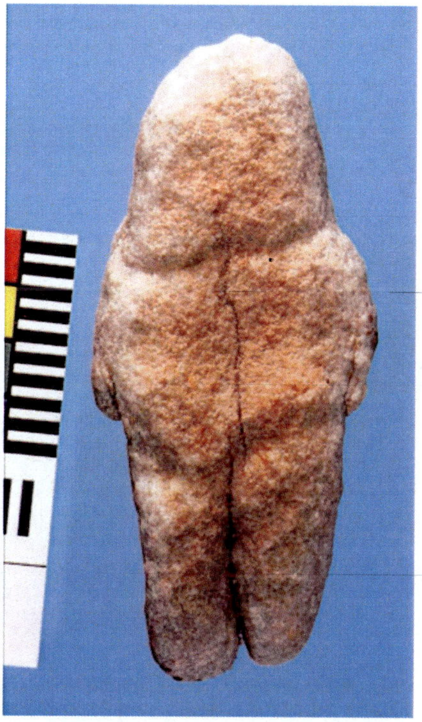

Fig. 3.10 Middle Acheulian proto-figurine, Tan-Tan, southern Morocco, most probably about 400,000 years old; scale in mm

been modified and microscopic traces of a red pigment seem to indicate that the figurine had once been coated by red paint.

Portable Engravings

The world's archeological community is largely unable to accept any evidence of early symboling, and it, therefore, remains divided over the status of the several engraved objects reported from the Lower Paleolithic. The largest site assemblage is from Bilzingsleben, a major occupation site (more than 1000 m^2 excavated) of the Holstein Interglacial in Germany, about 300 ka old (Mania 1991). This biface-free industry of well over 100,000 stone tools has been found together with numerous very robust human remains. The lakeside living site has yielded six apparently engraved bone and ivory fragments, mostly of the forest elephant, and one marked quartzite slab (Mania and Mania 1988; Bednarik 1988, 1993c, 1995a). It is widely accepted that the grooves found on these specimens were made with the points of stone tools, but some commentators have considered them to be incidental results of utilitarian activities. However, the D-shaped marking on the stone slab shows repeated application of a tool to master its difficult curved part. While most of the other engravings are merely groups of linear grooves, those on the first four bone objects reported have been demonstrated by lasermicroscopic analysis to have been made intentionally (Steguweit 1999). Five of the bundled subparallel grooves on bone object No. 3 (Fig. 3.11) were all made with the same stone tool (Bednarik 1988), and one of the three engraved Micoquian bone fragments from gravel pit

Fig. 3.11 Bilzingsleben engraved object No. 3, a forest elephant bone fragment with a set of convergent lines, Germany, of the Holstein Complex; scale in cm

Fig. 3.12 *Top*: Bilzingsleben engraved object No. 1, fragment of a forest elephant tibia with two sets of engravings, Germany, c. 300,000 years old. *Bottom*: Oldisleben engraved object No. 1, Germany, in the order of 100,000–130,000 years old

Oldisleben 1, Thuringia (Germany) (Bednarik 2006a), displays markings almost identical to those on the No. 1 object from Bilzingsleben (Fig. 3.12).

The status of a similarly marked elephant bone from another central European hominin site, Stránská skála in the Czech Republic (Valoch 1987), remains to be clarified, although it does resemble the marking strategies of other very early finds. The lines on a fragment of an ox rib, Acheulian, Pech de l'Azé, France (Bordes 1969; Marshack 1977), are judged here to be natural phenomena. The status of two marked bone fragments, one a bovid bone 8 cm long with about ten grooves, the other a cervid bone fragment bearing twenty-seven notches along an edge, from Kozarnika Cave, in northwestern Bulgaria, still needs to be established. Found in early 2004, together with many proto-Levallois stone implements and small bifacial forms, this has been attributed to the Early Pleistocene by paleomagnetism and claimed to be 1.1–1.4 million years old. Similarly, the nature of the various markings near the tip of a Stegodon tusk, deposited together with a second tusk from a different animal (Gao et al. 2004) in Xinglongdong Cave, China, need to be verified. Some do appear to be intentional, and they are from an occupation layer with a human tooth, between 120 and 150 ka old.

By contrast, the anthropic authenticity of an engraved bone fragment from the Acheulian of Sainte Anne I, France, which bears ten short cuts along an edge, seems assured (Raynal and Séguy 1986; Crémades 1996). This probable horse bone from near Polignac in the Haute-Loire region is remarkably similar to the German fragment of a mammoth tusk from Whylen near Lörrach. The latter bears a series of about twenty short, obliquely cut notches, arranged linearly and so evenly spaced that they seem to be notational (Moog 1939). The age of the ivory fragment is not known but as it was excavated in a Rissian loess it is probably of similar antiquity as the French specimen, belonging to the late Lower Paleolithic. The German specimen was lost at the end of World War II and has probably been "souveniered." A bone fragment bearing incised lines comes from Stratum 5BS at Border Cave, with ESR ages of about 170 ka, but with micromammal data indicating a prior, OIS 7 ascription (Grün and Beaumont 2001), at 240–190 ka ago (Martinson et al. 1987). Finally, Wonderwerk Cave in South Africa has yielded a fragment of silicified "ironstone" bearing a set of slightly curved, subparallel lines incised with a stone tool. It is from a late Fauresmith context, dated to between 280 and 276 ka (Imbrie et al. 1984; Bednarik and Beaumont 2010), and thus of an antiquity matching that of the Bilzingsleben finds in order of magnitude (Fig. 3.13). In two of the incised grooves, which on this very hard material would have been produced with difficulty, the tool was raised briefly before continuing its course, emphasizing the intentionality of the markings (Bednarik and Beaumont 2010).

With about a dozen credible specimens at our disposal, the case for Lower Paleolithic engravings on portable objects remains tenuous. However, the consistencies among these finds, particularly in the marking strategies employed (Bednarik 1995a; Hodgson 2000), as well as their continuation in the far more numerous specimens from the subsequent Mode 3 industries (Bednarik 2003b), demand their serious consideration. There are some distinctive patterns: the markings, clearly made with stone tools, appear to be responses to the shape of the available facet area in most instances. Only two of the marking sets seem to be randomly arranged. Nearly all of them show apparently deliberate spacing of individual marks (cf. Bednarik 2006a), and other indications of purposefulness are present. Bearing

Fig. 3.13 Engravings of the Fauresmith tool tradition, Wonderwerk Cave, South Africa, between 276,000 and 280,000 years old; scale in mm

in mind that the use of coloring material is safely demonstrated from the late Lower Paleolithic, and that in some cases crayons were used to mark rock surfaces by stroking, it should be a reasonable expectation that such marking of surfaces was also attempted by abrasive or cutting action. Such action was widely used in the utilitarian technology of the period, as we know from its wooden artifacts.

Beads and Pendants

It is well known that Jacques Boucher de Crèvecœur de Perthes (1788–1868) was one of the first to demonstrate the existence of the Paleolithic. (John Frere, a direct ancestor of Mary Leakey, is credited with having been the first to propose this in 1797 [Leakey 1984: 15].) But it was soon forgotten that with the "handaxes" and animal remains he and Marcel-Jérôme Rigollot excavated at Abbeville and St Acheul, they also found a large number of fossilized sponge fragments with central perforations (named as *Coscinopora globularis*), which may or may not be manuports (Boucher de Perthes 1847). Rigollot (1854) considered them to have been used as beads (Prestwich 1859: 52), while Prestwich himself, who also found some specimens, remained undecided but did note that some of the holes appeared to have been enlarged artificially. Because the pieces found no further attention, they had been forgotten by the time Smith (1894: 272–276) excavated about 200 identical items from an Acheulian site at Bedford, England. These were of precisely the same species and also showed artificial enlargement of the natural orifice. Smith was certain that these specimens were used as beads, which in view of the identical French finds, from St Acheul, Amiens, Soissons, Le Pecqu and Paris and of roughly the same period, is indeed likely (see also Thieullen 1901: figure 4). Keeley (1980: 164) examined some of the English sample and confirmed that there is no doubt that their perforations were modified. Goren-Inbar et al. (1991) recovered similar disc-like and perforated fossil casts from an Acheulian site, Gesher Ya'aqov in Israel, although these are crinoid segments (*Millericrinus* sp.) and no evidence of modification was noted. A thorough microscopic examination of 325 of the French and English globular specimens conducted in 2003 found that many of these were indeed significantly modified. Moreover, many of them also bore distinctive wear facets around their openings, indicating that they were worn over a long period of time while threaded on a string. Therefore, their status as beads should be regarded as reliably demonstrated (Fig. 3.14). However, they had been incorrectly identified since the 1840s; these objects are without exception of the species *Porosphaera globularis* Phillips 1829, a Cretaceous sponge (Bednarik 2005).

What renders this finding even more plausible is the discovery of clearly made disc beads from a Late Acheulian site in Libya, El Greifa (Ziegert 1995; Bednarik 1997). According to Th/U dating and other evidence, these ostrich eggshell beads are about 200 ka old, and replication experiments have shown that their manufacture involved a complex procedure (Fig. 3.15). Originally, only three damaged specimens were found, but about forty more have become available since (Ziegert 2007). Furthermore, there can be no doubt about the authenticity of two pendants

Fig. 3.14 Six Acheulian beads made from *Porosphaera globularis* fossils. Note the very heavy wear that resulted in a distinct wedge shape of (**b**), the thin centric wear facets on (**a**) and (**c**), and the light-colored, distinctly asymmetric major wear facet on (**d**). Specimen (**e**) is fractured, and (**f**) shows very little use wear

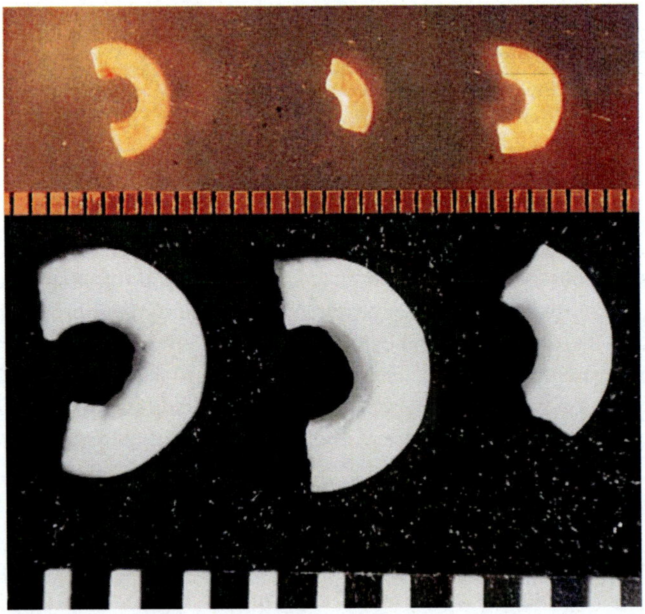

Fig. 3.15 *Top*: three Acheulian ostrich eggshell beads from El Greifa, Libya, c. 200,000 years old (photograph by Helmut Ziegert, reproduced with permission). *Bottom*: experimental replicates produced by the author; scale in mm

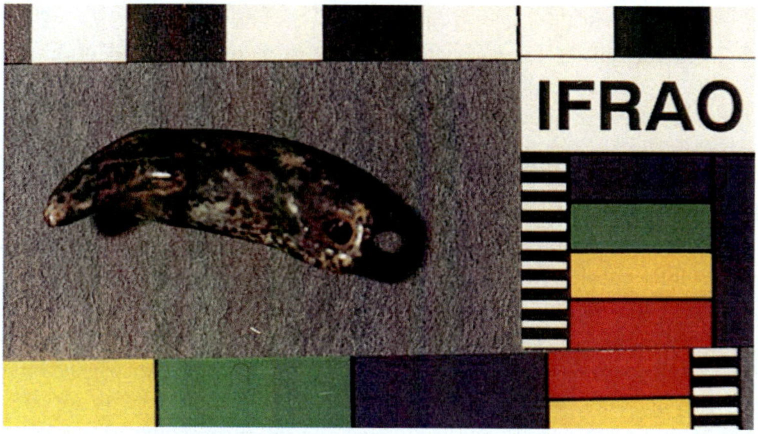

Fig. 3.16 Pendant from the Repolusthöhle in Austria, wolf incisor, perforated near its root; probably late Lower Paleolithic

from the Repolust Cave in Styria, Austria (Bednarik 1992, 1997), Rigaud et al. (2009) notwithstanding. Their perforations are clearly anthropic, but since their discovery (Mottl 1951) they have attracted almost no attention. A drilled wolf incisor (Fig. 3.16) and flaked bone point were recovered together with a large lithic assemblage variously described as Levalloisian, Tayacian and Clactonian, probably a late Lower Paleolithic industry. It was found well below an Aurignacian or, more likely, Olschewian level, separated from it by substantial clastic deposits of stadial periods. No reliable dating is available from the site, but according to the regionally well-known paleontology, especially the phylogeny of the bears, the Mode 1 occupation seems to be in the order of 300 ka old.

Two Incommensurable Models

The incommensurability of the short-range (replacement, discontinuist or African Eve; henceforth SR) model and the long-range (gradualist; henceforth LR) model (d'Errico and Nowell 2000), separating two mutually exclusive explanations of hominin evolution, is well illustrated by these last-mentioned examples. To preserve their beleaguered hypothesis the SR protagonists have no choice but to reject the authenticity, dating, or attribution of any find that suggests advanced cognitive abilities in hominins prior to the advent of the "Upper Paleolithic." Not only are they obliged to discredit literally all of the empirical evidence listed above in this chapter, they also have to find ways of negating all paleoart evidence of what they call Middle Paleolithic periods, Middle Stone Age, or more correctly, Mode 3 industries. Put simply, this is an impossible task. For instance, all of the Pleistocene and early Holocene rock art of Australia, a huge corpus of evidence numbering hundreds of thousands of motifs, is clearly and indisputably of Mode 3 production (Bednarik

2010a, b). So is all of the rock art of Tasmania, and all of the paleoart of Africa up to perhaps 30 ka ago. In other words, a massive body of "art" production is the creation of Mode 3 technological traditions. There is in fact far more surviving "Middle Paleolithic" rock art in the world then "Upper Paleolithic," which renders the claims that "modern" behavior patterns were introduced with the latter technological phase completely farcical.

The Sisyphean task of the SR advocates is correspondingly hopeless, and their feeble endeavors to find ways of rejecting possibly millions of instances of paleoart predating Mode 4 technologies are most illuminating. This fact alone renders the SR notion absurd and indefensible. Let us be perfectly clear on this issue: somewhere between 90 and 99% of all the world's rock art that has managed to survive from the Pleistocene period was created by people with a Mode 3 or earlier rather than a Mode 4 technology. Most of the world's Pleistocene archeologists are completely unaware of this, which surely questions their competence in making any pronouncements about the subject. The SR model demands that symboling and any "modern" behavior modes are exclusive to Mode 4 traditions. Hence the SR model is severely impaired before we even consider its more detailed claims. It is an academic farce, pure and simple; it was proposed and then vigorously defended by people whose relevant knowledge was arguably inadequate. The epistemological ineffectualness of their program is illustrated by the dialectic they apply to this, which is usually marked by an initial lack of knowledge about the existence of evidence contradicting their view. When this is pointed out, rather than conceding their shortcomings they respond by attacking the data they were unaware of, or the credibility of the presenters. Often disguised as refutation attempts, this pattern of response is evident in many examples. Here are some of them analyzed, beginning with the last-listed beads and pendants:

1. Such "Lower Paleolithic" small, perforated objects have been rejected by SR advocates by disputing their anthropic status or their chronological attribution. The fossil sponge beads, it has been claimed, are not modified and the signs of wear *may be* natural (Rigaud 2006–2007). However, to demonstrate the anthropic nature of these assemblages of fossil casts does not rely on these aspects. Only about one-quarter of a natural sample of the *Porosphaera globularis* casts is of reasonably spherical shape; only 14% of them feature the natural tunnels; and less than 10% fall between 10 and 18 mm diameter, the size range of most of the beads. Therefore, less than 0.3% of a random sample would be of the shape and size range of the collections of beads from Acheulian sites, and *also* have some degree of tunneling (Bednarik 2005). Moreover, not a single undamaged specimen would have a tunnel with *two* openings, as all of the beads obviously do. What the SR proponents fail to understand is that these tunnels are the result of a predator-avoidance strategy by the sipunculan worm *Trypanites weisei* Mägdefrau 1932 or a similar species. That means that, although they sometimes come close to breaking through at the fossil's other side, they never actually do (Neumann et al. 2008). Creating a second tunnel entrance would defeat the worms' protection strategy; therefore, there are no undamaged natural

specimens known with two openings of the same tunnel. This renders it virtually impossible to explain collections of specimens from Acheulian sites that are of a specific size range and shape, exclusively with tunnels broken through, by any means other than intentional collection. No natural process could possibly account for such occurrences. Therefore, the modifications observed on these hundreds of collected fossils are not even needed to invoke anthropic agency. By focusing their attention on attempting to refute the modification and wear traces, the SR advocates have only demonstrated that they did not understand the biological data and instead attempted a redundant falsification. This example illustrates the accommodative pattern of reasoning of the SR people: they are not interested in what happened in the human past; their motivation is to defend the dogma and to retreat from that position as reluctantly as possible (cf. Rigaud et al. 2009). It also illustrates the practices involved: although presenting precisely the same data as Rigaud (2006–2007), Rigaud et al. (2009) document a significant retreat from the position expounded in the older version, and yet they fail to cite the original position, thus concealing the tergiversation.

2. Language is a form of symboling, and since the SR scholars cannot admit the systematic use of symbols prior to the creative explosion they believe occurred around 40 ka ago, there have been almost countless attempts to demonstrate that "reflective verbal communication" first appeared at that point in time. For instance, it has been proposed that iconographic art had to precede language (Davidson and Noble 1989), and that no evidence of it is available from Mode 3 traditions. This claim was incorrect when it was made (the painted zoomorphs from Apollo 11 Cave are of the MSA; Wendt 1974; although that claim is itself now refuted), and today we even have an apparently iconic motif from the Micoquian (Bednarik 2006a). Discussions of language ability have focused on Broca's and Wernicke's areas, which have both been claimed to be detectable on cranial endocasts of *Homo habilis*. They may not necessarily be reliable indicators of language ability, but their very early presence may indicate that some of the required structures were available to habilines. Such structures can only evolve if selective pressures favor them, so selection in favor of speech must have preceded them. Much of the language origins debate has focused on the hyoid Neanderthal bone from Kebara Cave (Arensburg et al. 1989; Marshall 1989; notwithstanding Lieberman's 2007 speculations) which only indicates their unproductive and even irrelevant nature. The Dikika infant's hyoid bone (Alemseged et al. 2006) renders these discussions entirely superfluous: it shows poignantly how the historical sequence and occurrence of finds determines the profound transience of our constructs of the past. The major recent syntheses on the subject tend to return to linguistic and archeological perspectives (Bickerton 1996, 2010; Dunbar 1996; Aitchison 1996; Falk 2009), and their authors arrive at the same basic finding: human language is such a complex phenomenon that its evolution, in every sense, must have been a very lengthy process. It cannot possibly be accommodated in the SR model, whose advocates were consistently unaware of the precariousness of their ideas in that respect.

3. Similarly, the SR lobby needs to disproof the occurrence of interment practices prior to the purported cognitive quantum jump, because they would challenge its credibility. Gargett's (1989, 1999) literature-based reviews squarely reject the notion of "Neanderthal" burials, but the evidence that at least some of these Robusts were deliberately buried is simply overwhelming (Harrold 1980; Bar-Yosef 1988; Smirnov 1989; Defleur 1993; Otte 1996) and it is widely accepted today. Again, the opposition to early interment practices was always attributable to inadequate knowledge of the presented data, much of which has been available since the first half of the twentieth century (e.g., from La Ferrassie in France).
4. Rock art predating Mode 4 traditions discredits the SR model, yet it is very common, especially in Australia and southern Africa. But if it predates Mode 3 industries, as it does in a very few Indian and African cases, it is complete anathema to that model. Yet the domination of very early rock art by these cupules is very probably a taphonomic phenomenon, therefore it tells us not much about these paleoart traditions or their range of expressions. Nevertheless, they are important to the origins of symboling because there can be no question about either their intentionality or their semiotic nature. The manufacture of cupules was highly labor intensive and they have no utilitarian roles whatsoever. So far SR scholars have hardly responded to the reports of "Lower Paleolithic" petroglyphs, and it remains to be seen how they will try to explain away this evidence. Many of their most recent discussion articles (e.g., Rigaud et al. 2009) imply that they may be unaware of it, which would again confirm the familiar pattern.
5. Mineral pigments such as hematite, goethite, and ochreous materials could conceivably be used for utilitarian purposes, although this is not common ethnographically and unlikely for the "Lower Paleolithic." The likelihood that these pigments were used for symboling activities (body painting, coloring of artifacts, coloring of rock surfaces) or to "make special" (sensu Dissanayake 1988) is much greater. In some cases it has been demonstrated that the "Acheulians" applied crayon-like hematite pieces to rock surfaces. It is odd that, of the various types of evidence suggesting very early symboling, pigment use is the only one that has been widely accepted by orthodox archeology, despite being the perhaps most ambiguous of all these types. Nevertheless, the SR lobby has not so far presented a lucid manifesto of how it explains the "Lower Paleolithic" use of pigments within the framework of its model. In this case it is impossible that the main protagonists are unaware of this evidence, so perhaps they could explain what people without symboling ability used these pigments for.
6. Whenever confronted by individual finds that challenged the SR model, its defenders have traditionally and consistently tried to explain them away, or regarded them as a "running ahead of time" (Vishnyatsky 1994), or pronounced them as untypical and rare, or challenged their dating or the scholarly competence of their promoters. This is a familiar pattern in Pleistocene archeology, dating back to the times of Boucher de Perthes and Pengelly, up to the mid-1800s—the "incompetent amateurs" who discovered the Paleolithic—as well as to the later, similarly "incompetent" discoverers of fossil man (Fuhlrott), Pleistocene art (de Sautuola) and *Homo erectus* (Dubois), and many more scholars since.

Unfortunately this pattern of denial will continue—being attributable to a defense system Pleistocene archeology has developed because of the impossibility of subjecting most of its claims to systematic refutation. An extreme conservatism has led to a reliance on, and preference for, authority, which is already an epistemological impairment. But more relevantly, it fostered a specific brand of minimalist dogmas. These are based on the assumption that to protect the paradigm against unsound challenges to what provides a "semblance of certainty," it is best to resist changes to a dogma. The more we resist, the closer the process resembles an inverted form of falsificationism. In other words, if according to the dogma humans acquired a specific technology or ability at a certain time, any notion supporting an earlier introduction has to be resisted strenuously, until the evidence becomes simply overwhelming. This protects the received knowledge against frivolous claims, whereas those that do not challenge the dogma could be readily accepted.

This shows that anything can be proposed and will be accepted, provided it leaves the dogma intact: *compliance with dogma, not veracity, is the criterion of acceptance*. But therein lies the problem: the dogma is itself likely to be false, and this provides a key to understanding the false epistemology of Pleistocene archeology. To see this, we need to appreciate how the discipline is entirely dependent upon a random historical sequence of discoveries: if that sequence had been different, our received knowledge would be so also. Pleistocene archeology is entirely at the mercy of the historical sequence in which key discoveries are made—those that guide the dominant paradigms. In contrast to the systems of data gathering in most other disciplines, there can be little design in the knowledge acquisition strategies of Pleistocene archeology. Most key finds are made fortuitously, yet they may decide how other aspects are interpreted. For instance, when the period's first rock art was reported, from Altamira in Spain, it was completely rejected for decades. Its sophistication was considered entirely incompatible with the perceived primitiveness of Upper Paleolithic people, as deduced from their earlier found tools (which, incidentally, had also been categorically rejected earlier). Yet it is obvious that if Paleolithic cave art had been discovered, reported, and accepted first, it would have been the tools that would have been rejected as being contemporary, because *they* would have been regarded as incompatible with the sophistication of the art. In either case, the perceptions and expectations of scholars will be significantly distorted, yet we have no control over the order in which discoveries are made and accepted. Similarly, their acceptance depends on perfectly subjective factors at any time, and that certainly has been the case since the nineteenth century and *has continued right to the present*.

If the Pleistocene archeology we believe in at any point in time would be different had the sequence and availability of discoveries been different, the dogma is clearly a reflection of random factors and contingent prejudices. To defend a randomly acquired model on no basis other than its historical precedence is demonstrably fallacious, otherwise we would have to prefer astrology to astronomy, phrenology to the cognitive sciences. Moreover, the practice of retreating as reluctantly as possible from such an incomplete model is logically unsupportable. It argues for a regression from a *contingent* state of limited validity, instead of one constructed on falsifiable propositions. This is epistemologically obvious, and yet it is this very practice that determines archeological dogma, which then has to be preserved at the cost of discarding or discrediting valid evidence. It is this inverted falsificationism that is the very reason why only one thing is certain about any archeological model: it *must* be inherently false and it *will* be replaced by another in due course. The only workable solution to this endless succession of contingent speculations is to replace them with the null hypothesis that sampling errors are inherent in all archeological work. Then we can step back, look at the whole picture, and understand the severe limitations and fragility of our knowledge base. We can then perhaps even replace archeology's false taxonomies, the circular reasoning, and the authority-based pronouncements of the past with a universal theory.

Toward a Universal Theory

The probably greatest single epistemological encumbrance of archeology as it has been conducted is the tendency of treating "empirical evidence" as representing a random sample—as if it amounted to a representative selection of variables defining the entity being explored. The concept of "random sample" is taken from the practices of the hard sciences, where it is crucial that sampling units are in fact representative. In archeology, however, it is impossible to secure samples of culture that can be representative of any condition: each site deposit, and each part of each site, is unique. Representativeness is manufactured by the archeologist, who arranges series of objects arbitrarily and creates a taxonomy from them, then attributing them to "cultures." To illustrate how this process can lead to absurd systems, we may consider its application to rock art. Major rock art sites are almost always cumulative assemblages in generally two-dimensional space. The scientific dating of these sequences remains extremely difficult (Bednarik 2002b). So we have single sites or rock panels bearing the artistic precipitate of different periods, perhaps different cultures. Orthodox archeology then "names" motifs, creates taxonomies, and "discovers" traditions that are completely fictional.

Obviously this can only lead to falsities, and there are countless examples in archeology where the same errors of logic occur. In epistemological parlance, there is a dependency relation called a supervenience: one set of properties (forming a historical event) is supervenient on a second set (represented in the selected sample). The relationship between the two sets cannot, therefore, be explored by traditional deductive reasoning. However, even if one made allowances for the purely taphonomic issues (the enormous variations in the survival rates of different classes of evidence), the disparities would not be solved. These variations are much greater than most practitioners realize. Of all the events that occurred during the archeological past, no evidence of any kind survived for more than a second in 99.999% of all cases. Of the still innumerable remaining instances, evidence survives to this day only in a tiny fraction of one-millionth of a percent. Of this remaining "sample," only an infinitesimal portion can reasonably be assumed to have been recovered, of which an even smaller part has been correctly interpreted. This introduces an even more profound issue: not only do we need to understand the systematic biases of preservation; we also need to consider those of recovery and interpretation, and the systematic dynamics they engender.

The observation that a "taphonomic logic"-style of discourse (Bednarik 1994a) needs to be applied to various factors other than taphonomy proper expands the scope of this discussion considerably. The solution lies in the introduction of *metamorphology* as the scientific version of archeology. It is a logic-based, refutable system of reviewing archeological information that determines whether archeological propositions could have scientific legitimacy. It is developed especially from taphonomic logic, which hinges on the concept of cumulative data loss as a function of time (the principle is depicted graphically in Fig. 3.17). It replaces inductive uniformitarianism, hitherto the de-facto basis (Cameron 1993), as a unified theory of archeology.

Metamorphology (Bednarik 1995b, 2006b) is the science of how forms of evidence of events in the past become the forms as which they are perceived or understood by the individual researcher today. In accounting for the considerable gap that exists between the reality of what actually happened at some point of time in the distant past, and the abstraction of it as it is perceived by, for example, an archeologist, it is crucial to focus on the interpreting individual. The discipline is not a quasi-democratic reflection of the view of all; its paradigm is based on the authority of a political hegemony, and is interpreted by the individual. Metamorphology obviously has to take into consideration myriad factors and it cannot be expected to provide precise interpretations, but it needs to determine how the individual interpreter of the past arrives at his or her pronouncements. Knowledge in archeology is not some mysterious collective unconscious to which practitioners are somehow

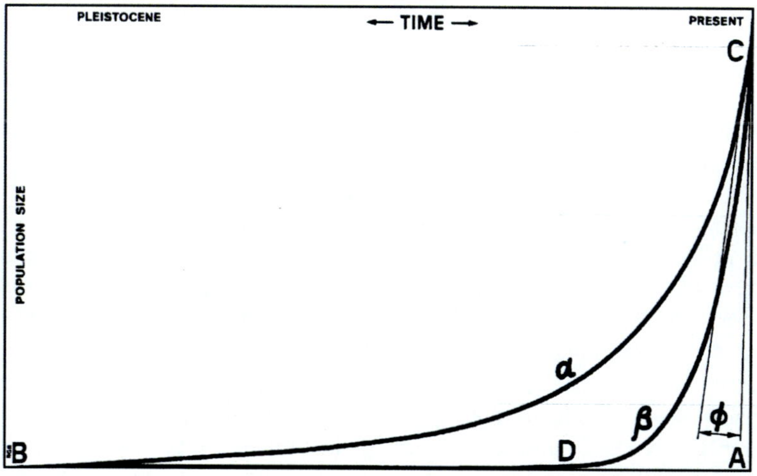

Fig. 3.17 The principles of taphonomic logic: the relationship of total production of an archeological phenomenon s_α to its surviving instances s_β as a function of angle ϕ

connected; it is individual knowledge of individual practitioners, limited by many factors. Like refutation in general, metamorphology provides us with models of what is unlikely to be valid, and so strengthens archeology by weakening its dogmas. It also rejects the concepts of an "archeological record" and a collective knowledge of the discipline.

The most obvious of these factors accounting for metamorphology, and for the gap between archeologists' constructs and what really happened in the past, is taphonomy. It distorts archeological evidence systematically, and it does so in forms that have often not been appreciated adequately. Indeed, after the paleontological concept of taphonomy was introduced into archeology over 40 years after its inception (Efremov 1940), it was soon misunderstood and was effectively applied in the manner of actuopaleontology—which ironically taphonomy was originally intended to replace (see Solomon 1990 for a superb discussion). Hence the potential of taphonomy itself has remained significantly under-utilized in archeology in more ways than one. But if the inherent principles are extended to the methods of recovery of evidence—those of its interpretation; those of its reporting and selective dissemination; those of its statistical treatment; or to the individual researcher's own biases and limitations, such as limitations of knowledge or language; and to a variety of other factors—it becomes apparent that these also tend to be *systematic*. These factors may include the priorities of research traditions, of individual leaders in the discipline, of specific institutions, of funding agencies, or of society as a whole. There can be no doubt that there is a very considerable gap between the reality of what happened in the distant past and the abstraction of it as perceived by the individual archeologist interpreting a specific, subjectively selected, and nonrandom sample of the remaining evidence. Taphonomy itself is not the whole answer, because it accounts for only *some* of these truncating and modifying factors.

For metamorphology to be scientific, its propositions must be refutable. It is logic based and draws heavily on the understanding of taphonomic processes, and on a variety of other falsifiable observations (Fig. 3.18). A unified theory of metamorphology has been formulated, at least in embryonic form (Bednarik 1995b, 2006b). It has been shown that metamorphological quantification, although extremely difficult, should be possible, at least in general or abstract forms (i.e., as integral functions). It extends the underlying principle of taphonomic logic (that scientific access to the human past is contingent on the coherent identification of *that part of the extant characteristics of the evidence that is not the result of taphonomic processes*; Bednarik 1990–91) to all aspects of archeological interpretation. These include the way data are *collected, stored, interpreted, and disseminated*. They include the biases of the individual researcher (cognitive, religious, ontological, academic, intellectual), of specific schools or the discipline as a whole, and many other external factors that have a bearing on how the so-called empirical evidence is individually perceived, reported, and interpreted. For instance, the researcher's own limitations are a powerful factor in how evidence may be reported. These may be limitations of knowledge or of language. Ignorance of researchers concerning existing data, language barriers, and biases through preconceived models have not only severely influenced hypotheses and their defense; they have also

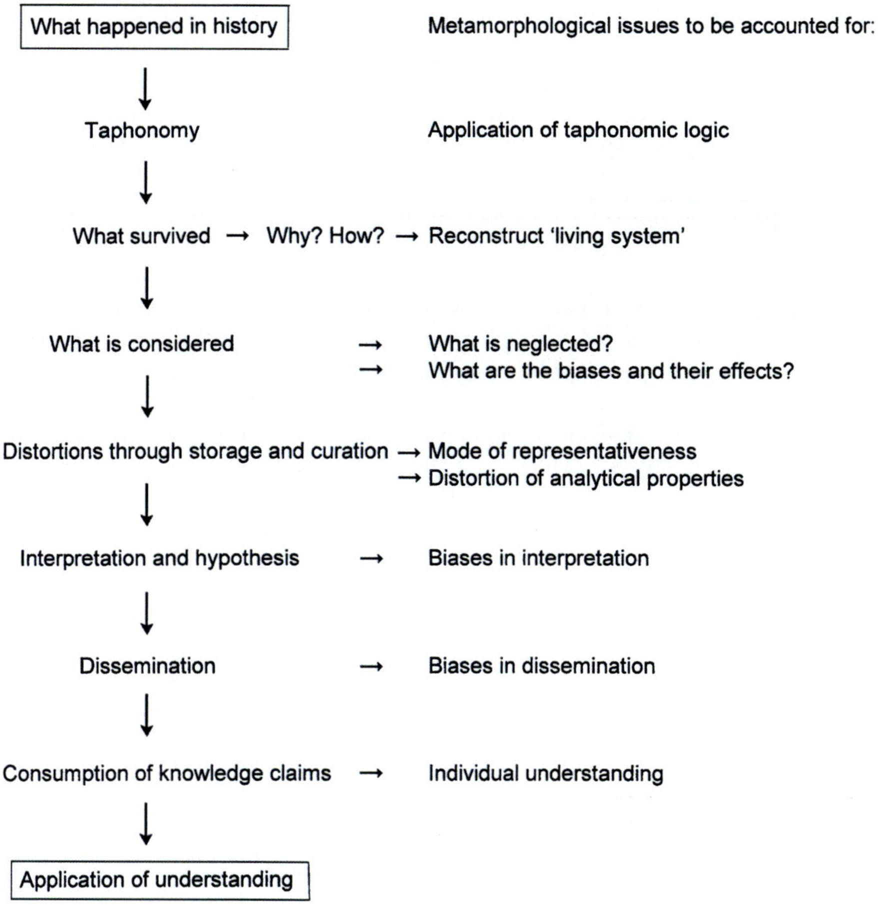

Fig. 3.18 Principal factors in metamorphological analysis of archeological knowledge claims

stifled the flow of information in paleoart studies and archeology (see e.g., Bednarik 1992, 1995b, c, 1999b). It is certainly a quantifiable factor. The academic system itself, which is so crucial to the dissemination of knowledge, can also stifle that very process and act as a filter in quite a number of ways. The term *déformation professionnelle* refers to this issue: professional training also results in a distortion in the way the world is perceived. Confirmation bias (Wason 1960; Evans et al. 1983) can only add to the sophistry. Observer-relative definitions, attributions, and claims about the distant human past are clearly not in themselves of scientific utility; they need to be subjected to metamorphological analysis, which so far has not occurred in a systematic fashion. All of this can cumulatively add up to such distortions in dominant models that these bear little resemblance to what historically happened in the past. This is because most of the distortions are not random; *they are systematic*.

To correct this we need to be able to understand the nature and effects of these distortions, be they taphonomic or related to other epistemic encumbrances. This would provide the kind of framework we require to account for the gap between what happened in the distant past, and the abstraction or reified construct of it as it is perceived by the individual researcher interpreting a specific "sample" of the remaining evidence of this event, or connected events. For instance, we need to understand the effects of false hypotheses and of their ardent defense if we are to obtain a valid reflection of metamorphology.

This applies particularly to hypotheses about the cultural dimensions of Pleistocene hominins, which, when subjected to metamorphological probing, fail decisively. The hard evidence presented in this chapter demonstrates that several classes of cultural (rather than technological) evidence cannot be realistically explained as anything other than the traces of symboling. The publications of the SR advocates show that, for the greater part, they were unaware of this huge corpus of evidence, and that where they do acknowledge the existence of it, they either ignore the impact it has on their ideas (as in the case of pigment use) or try to explain it away or reject its significance. Their defense of a minimalist dogma based on an archeologically, paleoanthropologically, and genetically highly unlikely scenario is historically superseded and based on inadequate knowledge. The dogma of orthodox Pleistocene archeology is a reflection of historical random factors and the contingent prejudices deriving from the uncertainties of an unfalsifiable discipline. Their mantra that people of Mode 3 or earlier technologies lacked symboling capacities, including language, stands completely refuted and discredited. Their responses to the presentation of the refuting evidence are uniformly unsatisfactory: in no case has there been an admission that the data base is significantly greater than had been assumed by the SR protagonists (see for instance, the debates of Chase and Dibble 1987; Davidson and Noble 1989; Bednarik 1992; Chase and Dibble 1992; Davidson 1992; or the equally fierce debate following Bednarik 1995a). They seem to have considered their hypothesis of a cognitive explosion on the first day of the Upper Paleolithic unassailable, and yet it does not even need to be refuted: its falsity is so perfectly transparent it should not even need to be demonstrated. With the world's massive corpus of Mode 3 rock art, the lack of an early African Late Stone Age and the seafaring prowess of hominins up to a million years ago, the notion of a cataclysmic event in Europe about 40 ka ago is a complete absurdity. This has all been known for many decades, as has most of the long list of symboling evidence from the "Lower Paleolithic" above, and the vastly greater amount of such evidence from the "Middle Paleolithic." The rise of the SR model is the result of inadequate knowledge of the archeologists who promoted it, and of those who all too eagerly embraced it. It is as severe an indictment of the epistemology of Pleistocene archeology as is the influence of the African Eve notion, introduced by an academic charlatan and eagerly adopted by the discipline.

To correct such an unsatisfactory state we need to be able to understand the nature and effects of these distortions. We need a kind of framework to account for the gap between what happened in the distant past, and the abstraction or reified construct of it as it is perceived by the individual researcher interpreting a specific "sample" of the remaining evidence of this event or connected events. Metamorphology provides such a framework. For instance, it can help us to understand the effects of false hypotheses and of their ardent defense. There is surely no reason why the dynamics of knowledge acquisition or academic power politics in the discipline should be immune from scholarly analysis. Archeology, like anthropology, often does not hesitate to study the taboos of the societies it investigates, be they extant (often to the chagrin of those "subjects") or extinct groups, so the study of itself should not be taboo either. These are realities; they have significant effects on the discipline, and these dynamics need to be understood like any other process contributing to our knowledge. Therefore, this aspect should be studied as carefully as any other that contributes to metamorphology. The discipline would be in a sorry state if such research would be discouraged because the "reputation" or sensitivities of individuals are considered to have precedence over its integrity or veracity—as demanded by some archeologists. Such intellectual corruption questions the value of the entire enterprise of Pleistocene archeology.

References

Aitchison, J. 1996. *The seeds of speech: language origin and evolution*. Cambridge University Press, Cambridge.

Alemseged, Z., F. Spoor, W. H. Kimbel, R. Bobe, D. Geraads, D. Reed and J. G. Wynn 2006. A juvenile early hominin skeleton from Dikika, Ethiopia. *Nature* 443: 296–301.

Arensburg, B., A. M. Tillier, B. Vandermeersch, H. Duday, L. Schepartz and Y. Rak 1989. A Middle Palaeolithic human hyoid bone. *Nature* 338: 758–760.

Bahn, P. G. 1997. Facing up to the earliest art. Paper presented to Symposium 3 of the Congreso Internacional de Arte Rupestre, Cochabamba, 2 April 1997.

Barham, L. S. 2002. Systematic pigment use in the Middle Pleistocene of south-central Africa. *Current Anthropology* 43: 181–190.

Bar-Yosef, O. 1988. Evidence for Middle Palaeolithic symbolic behaviour: a cautionary note. In M. Otte (ed), *Le homme de Néanderthal*, Vol. 5, *La Pensée*, pp. 11–16. ERAUL 32, Liège.

Beaumont, P. 1973. The ancient pigment mines of southern Africa. *South African Journal of Science* 69: 140–146.

Beaumont, P. 1990. Wonderwerk cave. In P. Beaumont and D. Morris (eds), *Guide to archaeological sites in the Northern Cape*. McGregor Museum, Kimberley, 101–134.

Beaumont, P. 1999. *Wonderwerk cave*. INQUA XV International Conference Field Guide: Northern Cape, pp. 27–31. University of Cape Town, Cape Town.

Beaumont, P. and R. G. Bednarik 2010. Pleistocene rock art from Africa. Paper presented to IFRAO Congress 2010, Foix.

Beaumont, P. and A. Boshier 1972. Mining in southern Africa and the emergence of modern man. *Optima*, March 1972 issue, 22(1): 2–12.

Beaumont, P. B. and D. Morris (eds) 1990. *Guide to archaeological sites in the Northern Cape*. McGregor Museum, Kimberley.

Bednarik, R. G. 1986. Comment on W. Davis, 'The origins of image making'. *Current Anthropology* 27: 202–203.

Bednarik, R. G. 1988. Comment on D. Mania and U. Mania, 'Deliberate engravings on bone artefacts of *Homo erectus*'. *Rock Art Research* 5(2): 96–100.

Bednarik, R. G 1990a. On the cognitive development of hominids. *Man and Environment* 15(2): 1–7.

Bednarik, R. G. 1990b. An Acheulian haematite pebble with striations. *Rock Art Research* 7: 75.

Bednarik, R. G. 1992. Palaeoart and archaeological myths. *Cambridge Archaeological Journal* 2(1): 27–43.
Bednarik, R. G. 1993a. Wonders of Wonderwerk Cave. *The Artefact* 16: 61.
Bednarik, R. G. 1993b. Palaeolithic art in India. *Man and Environment* 18(2): 33–40.
Bednarik, R. G. 1993c. Die Bilzingslebener Gravierungen im Lichte altpaläolithischer Beweise kognitiver Fähigkeit. *Ethnographisch-Archäologische Zeitschrift* 34: 550–554.
Bednarik, R. G. 1994a. A taphonomy of palaeoart. *Antiquity* 68: 68-74.
Bednarik, R. G. 1994b. Art origins. *Anthropos* 89: 169–180.
Bednarik, R. G. 1994c. The Pleistocene art of Asia. *Journal of World Prehistory* 8(4): 351–375.
Bednarik, R. G. 1995a. Concept-mediated marking in the Lower Palaeolithic. *Current Anthropology* 36: 605–634.
Bednarik, R. G. 1995b. Metamorphology: in lieu of uniformitarianism. *Oxford Journal of Archaeology* 14(2): 117–122.
Bednarik, R. G. 1995c. Wallace's barrier and the language barrier in archaeology. *Bulletin of the Archaeological Survey Association of Southern California* 20(3): 8–9, 13.
Bednarik, R. G. 1997. The role of Pleistocene beads in documenting hominid cognition. *Rock Art Research* 14: 27–43.
Bednarik, R. G. 1998. The australopithecine cobble from Makapansgat, South Africa. *South African Archaeological Bulletin* 53: 3–8.
Bednarik, R. G. 1999a. Pleistocene seafaring in the Mediterranean. *Anthropologie* 37(3): 275–282.
Bednarik, R. G. 1999b. Maritime navigation in the Lower and Middle Palaeolithic. *Comptes Rendus de l'Académie des Sciences Paris* 328: 559–563.
Bednarik, R. G. 2000. Early Indian petroglyphs and their global context. *Purakala* 11: 37–47.
Bednarik, R. G. 2001a. Replicating the first known sea travel by humans: the Lower Pleistocene crossing of Lombok Strait. *Human Evolution* 16(3–4): 229–242.
Bednarik, R. G. 2001b. An Acheulian figurine from Morocco. *Rock Art Research* 18: 115–116.
Bednarik, R. G. 2002a. An Acheulian palaeoart manuport from Morocco. *Rock Art Research* 19: 137–139.
Bednarik, R. G. 2002b. The dating of rock art: a critique. *Journal of Archaeological Science* 29(11): 1213–1233.
Bednarik, R. G. 2003a. A figurine from the African Acheulian. *Current Anthropology* 44: 403–413.
Bednarik, R. G. 2003b. The earliest evidence of palaeoart. *Rock Art Research* 20: 89–135.
Bednarik, R. G. 2003c. Seafaring in the Pleistocene. *Cambridge Archaeological Journal* 13: 41–66.
Bednarik, R. G. 2005. Middle Pleistocene beads and symbolism. *Anthropos* 100(2): 537–552.
Bednarik, R. G. 2006a. The Middle Palaeolithic engravings from Oldisleben, Germany. *Anthropologie* 44(2): 113–121.
Bednarik, R. G. 2006b. A unified theory for palaeoart studies. *Rock Art Research* 23: 85–88.
Bednarik, R. G. 2008. Cupules. *Rock Art Research* 25: 61–100.
Bednarik, R. G. 2010a. Pleistocene rock art in Australia. *Anthropos* 105(1): 3–12.
Bednarik, R. G. 2010b. Australian rock art of the Pleistocene. *Rock Art Research* 27(1): 95–120.
Bednarik R. G. and P. Beaumont 2010. Pleistocene engravings from Wonderwerk Cave, South Africa. Paper presented to IFRAO Congress 2010, Foix.
Bickerton, D. 1996. *Language and human behaviour*. UCL Press, London.
Bickerton, D. 2010. *Adam's tongue: how humans made language, how language made humans*. Hill and Wang, New York, NY.
Binneman, J. and P. Beaumont 1992. Use-wear analysis of two Acheulean handaxes from Wonderwerk Cave, Northern Cape. *South African Field Archaeology* 1: 92–97.
Bordes, F. 1969. Os percé mousterién et os gravé acheuléen du Pech de l'Azé II. *Quaternaria* 11: 1–5.
Boucher de Perthes, J. 1847. *Antiquités celtiques et antédiluviennes. Mémoire sur l'industrie et les arts a leur origine*. Treuttel et Wurtz, Paris.
Brecht, B. 1980. *The life of Galileo*. Methuen, London (German version 1939).

Cabe, P. A. 1980. Picture perception in nonhuman subjects. In M. A. Hagen (ed), *The perception of pictures*, Vol. II, pp. 305–343. Academic, New York, NY.
Cameron, D. W. 1993. The archaeology of Upper Palaeolithic art: aspects of uniformitarianism. *Rock Art Research* 10: 3–17.
Chase, P. G. and H. L. Dibble 1987. Middle Palaeolithic symbolism: a review of current evidence and interpretations. *Journal of Anthropological Archaeology* 6: 263–296.
Chase, P. G. and H. L. Dibble 1992. Scientific archaeology and the origins of symbolism: a reply to Bednarik. *Cambridge Archaeological Journal* 2(1): 43–51.
Clark, J. D. 1974. *Kalambo Falls prehistoric site*, Vol. 2, 436p. Cambridge University Press, Cambridge.
Clark, J. D., K. P. Oakley, L. H. Wells and J. A. C. McClelland 1947. New studies on Rhodesian man. *Journal of the Royal Anthropological Society* 77: 4–33.
Coss, R. G. 1985. Evolutionary restraints on learning: phylogenetic and synaptic interpretations. In N. M. Weinberger, J. L. McGauch and G. Lynch (eds), *Memory systems of the brain—animal and human cognitive processes*, pp. 253–273. The Guilford Press, New York, NY.
Crémades, M. 1996. L'expression graphique au paléolithique inférieur et moyen: l'exemple de l'Abri Suard (La Chaise-de-Vouthon, Charente). *Bulletin de la Société Préhistorique Française* 93(4): 494–501.
Crochet, J.-Y., J.-L. Welcomme, J. Ivorra, G. Ruffet, N. Boulbes, R. Capdevila, J. Claude, C. Firmat, G. Métais, J. Michaux and M. Pickford 2009. Une nouvelle faune de vertébrés continentaux, associée à des artifacts dans le Pléistocène inférieur de l'Hérault (Sud de la France), ver 1,57 Ma. *Comptes Rendus Palevol* 8: 725–736.
Dart, R. A. 1974. The waterworn australopithecine pebble of many faces from Makapansgat. *South African Journal of Science* 70: 167–169.
Davidson, I. 1990. Bilzingsleben and early marking. *Rock Art Research* 7: 52–56.
Davidson, I. 1992. There is no art—To find the mind's construction—In offence. *Cambridge Archaeological Journal* 2(1): 52–57.
Davidson, I. and W. Noble 1989. The archaeology of perception: traces of depiction and language. *Current Anthropology* 30(2): 125–155.
Davidson, I. and W. Noble 1998. Two views of language origins. *Cambridge Archaeological Journal* 8: 82–88.
Defleur, A. 1993. *Les sépultures moustériennes*. CNRS Editions, Paris.
de Lumley, H. 1966. Les fouilles de Terra Amata à Nice. Premiers résultats. *Bulletin du Musée d'Anthropologie Préhistorique de Monaco* 13: 29–51.
d'Errico, F., C. Gaillard and V. N. Misra 1989. Collection of non-utilitarian objects by *Homo erectus* in India. In *Hominidae. Proceedings of the 2nd International Congress of Human Paleontology*, pp. 237–239. Editoriale Jaca Book, Milan.
d'Errico, F. and A. Nowell 2000. A new look at the Berekhat Ram figurine: implications for the origins of symbolism. *Cambridge Archaeological Journal* 10: 123–167.
Dissanayake, E. 1988. *What is art for?* University of Washington Press, Seattle, WA.
Donald, M. 1991. *Origins of the modern mind: three stages in the evolution of culture and cognition*. Harvard University Press, Cambridge, MA.
Dunbar, R. 1996. *Grooming, gossip and the evolution of language*. Faber and Faber, London.
Efremov, J. A. 1940. Taphonomy: a new branch of paleontology. *Pan American Geologist* 74(2): 81–93.
Eitzman, W. I. 1958. Reminiscences of Makapansgat Limeworks and its bone-breccial layers. *South African Journal of Science* 54: 177–182.
Evans, J., J. L. Barston and P. Pollard 1983. On the conflict between logic and belief in syllogistic reasoning. *Memory and Cognition* 11: 295–306.
Falk, D. 2009. *Finding our tongues: mothers, infants and the origins of language*. Basic Books, New York, NY.
Feliks, J. 1998. The impact of fossils on the development of visual representation. *Rock Art Research* 15: 109–134.

Foley, R. and M. M. Lahr 1997. Mode 3 technologies and the evolution of modern humans. *Cambridge Archaeological Journal* 7: 3–36.
Gao, X., W. Huang, Z. Xu, Z. Ma and J. W. Olsen 2004. 120–150 ka human tooth and ivory engravings from Xinglongdong Cave, Three Gorges region, south China. *Chinese Science Bulletin* 49(2): 175–180.
Gargett, R. H. 1989. Grave shortcomings: the evidence for Neanderthal burial. *Current Anthropology* 30: 157–190.
Gargett, R. H. 1999. Middle Palaeolithic burial is not a dead issue: the view from Qafzeh, Saint-Césaire, Kebara, Amud and Dederiyeh. *Journal of Human Evolution* 37: 27–90.
Goody, J. 1977. *The domestication of the savage mind.* Cambridge University Press, Cambridge.
Goren-Inbar, N. 1985. The lithic assemblages of Berekhat Ram Acheulian site, Golan Heights. *Paléorient* 11: 7–28.
Goren-Inbar, N. 1986. A figurine from the Acheulian site of Berekhat Ram. *Mi'tekufat Ha'even* 19: 7–12.
Goren-Inbar, N., Z. Lewy and M. E. Kislev 1991. Bead-like fossils from an Acheulian occupation site, Israel. *Rock Art Research* 8: 133–136.
Goren-Inbar, N. and S. Peltz 1995. Additional remarks on the Berekhat Ram figurine. *Rock Art Research* 12: 131–132.
Gregory, R. L. 1970. *The intelligent eye*. Weidenfeld and Nicolson, London.
Grün, R. and P. Beaumont 2001. Border Cave revisited: a revised ESR chronology. *Journal of Human Evolution* 40: 467–482.
Harrold, F. 1980. A comparative analysis of Eurasian Palaeolithic burials. *World Archaeology* 12(2): 195–211.
Henshilwood, C. S., F. d'Errico, R. Yates, Z. Jacobs, C. Tribola, G. A. T. Duller, N. Mercier, J. C. Sealy, H. Valladas, I. Watts and A. G. Wintle 2002. Emergence of modern human behaviour: Middle Stone Age engravings from South Africa. *Science* 295: 1278–1280.
Henshilwood, C. S., J. C. Sealy, R. Yates, K. Cruz-Uribe, P. Goldberg, F. E. Grine, R. G. Klein, C. Poggenpoel, K. van Niekerk and I. Watts 2001. Blombos Cave, southern Cape, South Africa: preliminary report on the 1992-1999 excavations of the Middle Stone Age levels. *Journal of Archaeological Science* 28: 421–448.
Hodgson, D. 2000. Art, perception and information processing: an evolutionary perspective. *Rock Art Research* 17: 3–34.
Howell, F. C. 1966. Observations of the earlier phases of the European Lower Palaeolithic. *American Anthropologist* 68(2): 88–201.
Imbrie, J., J. D. Hays, D. G. Martinson, A. McIntyre, A. C. Mix, J. J. Marley, N. G. Pisias, W. L. Prell and N. J. Shackleton 1984. The orbital theory of Pleistocene climate: support from a revised chronology of the marine $\delta^{18}0$ record. In A. L. Berger, J. Imbrie, J. Hays, G. Kukla and B. Saltzman (eds), *Milankovitch and climate*, Part 1. Reidel, Dordrecht, 269–305.
Jacques, S. and P. D. Zelazo 2005. Language and the development of cognitive flexibility: implications for theory of mind. In J. W. Astington and J. A. Baird (eds), *Why language matters for theory of mind*, pp. 144–162. Oxford University Press, Toronto, ON.
Joulian, F. 1995. 'Human and non-human primates': des limites de genre bien problématiques en préhistoire. *Préhistoire Anthropologie Méditerranéennes* 4: 5–15.
Keeley, L. H. 1980. *Experimental determination of stone tool uses*, 60p. University of Chicago Press, Chicago.
Kumar, G. 1996. Daraki-Chattan: a Palaeolithic cupule site in India. *Rock Art Research* 13: 38–46.
Kumar, G., R. G. Bednarik, A. Watchman, R. G. Roberts, E. Lawson and C. Patterson 2003. 2002 progress report of the EIP Project. *Rock Art Research* 20: 70–71.
Laidler, P. W. 1933. Dating evidence concerning the Middle Stone Ages and a Capsio-Wilton culture, in the South-East Cape. *South African Journal of Science* 30: 530–542.
Leakey, L. S. B. 1958. Recent discoveries at Olduvai Gorge, Tanganyika. *Nature* 19: 1099–1103.

Leakey, M. D. 1971. *Olduvai Gorge. Vol. 3: excavations in Beds I and II, 1960–63*, pp. 306+xxi. Cambridge University Press, Cambridge.

Leakey, M. D. 1984. *Disclosing the past: an autobiography*. Doubleday, Garden City, NY.

Lieberman, P., 2007. The evolution of human speech: its anatomical and neural bases. *Current Anthropology* 48(1): 39–66.

Mania, D. 1991. The zonal division of the Lower Palaeolithic open-air site Bilzingsleben. *Anthropologie* 29: 17–24.

Mania, D. and U. Mania 1988. Deliberate engravings on bone artefacts of *Homo erectus*. *Rock Art Research* 5: 91–107.

Marshack, A. 1977. The meander as a system: the analysis and recognition of iconographic units in Upper Palaeolithic compositions. In P. J. Ucko (ed), *Form in indigenous art. Schematisation in the art of Aboriginal Australia and prehistoric Europe*, pp. 286–317. Australian Institute of Aboriginal Studies, Canberra.

Marshack, A. 1981. On Paleolithic ochre and the early uses of color and symbol. *Current Anthropology* 22: 188–191.

Marshack, A. 1996. A Middle Palaeolithic symbolic composition from the Golan Heights: the earliest known depictive image. *Current Anthropology* 37: 357–365.

Marshack, A. 1997. The Berekhat Ram figurine: a late Acheulian carving from the Middle East. *Antiquity* 71: 327–337.

Marshall, J. C. 1989. Reply to P. Lieberman, J. T. Laitman, J. S. Reidenberg, K. Landahl and P. J. Gannon, 'Folk physiology and talking hyoids'. *Nature* 342: 486–487.

Martinson, D. G., N. G. Pisias, J. D. Hays, T. C. Moore Jr. and N. J. Shackleton 1987. Age dating and the orbital theory of the ice ages: development of a high-resolution 0 to 300 000-year chronostratigraphy. *Quaternary Research* 27: 1–29.

McBrearty, S. 2001. The Middle Pleistocene of East Africa. In L. Barham and K. Robson-Brown (eds), *Human roots: Africa and Asia in the Middle Pleistocene*, pp. 81–92. Western Academic and Specialist Press, Bristol.

McFadden, P. L., A. Brock and T. C. Partridge 1979. Palaeomagnetism and the age of the Makapansgat hominid site. *Earth Planetary Science Letters* 44: 373–382.

McGrew, W. C. 1992. *Chimpanzee material culture: implications for human evolution*. Cambridge University Press, Cambridge.

McGrew, W. C. 1993. Brain, hands, and minds: puzzling incongruities in ape tool use. In A. Berthelet and J. Chavaillon (eds), *The use of tools by human and non-human primates*, pp. 143–157. Clarendon Press, Oxford.

Miller, G. H., P. B. Beaumont, A. S. Brooks, H. J. Deacon, P. E. Hare and A. J. T. Jull 1999. Earliest modern humans in South Africa dated by isoleucine epimerization in ostrich eggshell. *Quaternary Science Reviews* 18: 1537–1548.

Moog, F. 1939. Paläolithische Freilandstation im Älteren Löß von Wyhlen (Amt Lörrach). *Badische Fundberichte* 15: 36–52.

Mottl, M. 1951. Die Repolust-Höhle bei Peggau (Steiermark) und ihre eiszeitlichen Bewohner. *Archaeologica Austriaca* 8: 1–78.

Neumann, C., M. Wisshak and R. G. Bromley 2008. Boring a mobile domicile: an alternative to the conchicolous life habit. In M. Wisshak and L. Tapanila (eds), *Current developments in bioerosion*, pp. 307–328. Springer, Berlin and Heidelberg.

Noble, W. and I. Davidson 1996. *Human evolution, language and mind: a psychological and archaeological inquiry*, pp. 272+xiii. Cambridge University Press, Cambridge.

Nowell, A. 1995. Comment on R. G. Bednarik, 'Concept-mediated marking in the Lower Palaeolithic'. *Current Anthropology* 36: 621–622.

Oakley, K. P. 1981. Emergence of higher thought, 3.0–0.2 Ma B.P. *Philosophical Transactions of the Royal Society of London B* 292: 205–211.

Otte, M. 1996. *Le Paléolithique Inférieur et Moyen en Europe*. Armand Collin, Paris.

Pei, W. C. 1931. Notice of the discovery of quartz and other stone artifacts in the Lower Pleistocene hominid-bearing sediments of the Choukoutien Cave deposits. *Bulletin of the Geological Society of China* 11(2): 109–146.

Pelcin, A. 1994. A geological explanation for Berekhat Ram figurine. *Current Anthropology* 35: 674–675.
Pinker, S. 1997. *How the mind works*. Penguin, Harmondsworth.
Prestwich, J. 1859. On the occurrence of flint-implements, associated with the remains of extinct mammalia, on undisturbed beds of a late geological period. *Proceedings of the Royal Society of London* 10: 50–59.
Raynal, J.-P. and R. Séguy 1986. Os incisé acheuléen de Sainte-Anne 1 (Polignac, Haute-Loire). *RACF* 25: 79–80.
Rigaud, S. 2006–2007. Révision critique des *Porosphaera globularis* interprétées comme éléments de parure acheuléens. Unpublished MA thesis, Université Bordeaux 1.
Rigaud, S., F. d'Errico, M. Vanhaeren and C. Neumann 2009. Critical reassessment of putative Acheulean [sic] *Prosphaere globularis* beads. *Journal of Archaeological Science* 36: 25–34.
Rigollot, M.-J. 1854. *Mémoire sur les instruments en silex trouvés à Saint-Acheul, près Amiens*. Rapports géologiques et archéologiques, Amiens.
Smirnov, Y. A. 1989. Intentional human burial: Middle Palaeolithic (last glaciation) beginnings. *Journal of World Prehistory* 3: 199–233.
Smith, W.G. 1894. *Man the primeval savage. His haunts and relics from the hilltops of Bedfordshire to Blackwall*, 361p. Edward Stanford, London.
Solomon, S. 1990. What is this thing called taphonomy? In S. Solomon, I. Davidson and D. Watson (eds), *Problem solving in taphonomy: archaeological and palaeontological studies from Europe, Africa and Oceania*, pp. 25–33. Tempus, Archaeology and Material Culture Studies in Anthropology, Vol. 2. University of Queensland, St. Lucia
Sreenathan, M., V. R. Rao and R. G. Bednarik 2008. Palaeolithic cognitive inheritance in aesthetic behavior of the Jarawas of the Andaman Islands. *Anthropos* 103: 367–392.
Stapleton, P. and J. Hewitt 1928. Stone implements from Howieson's Poort near Grahamstown. *South African Journal of Science* 25: 399–409.
Steguweit, J. 1999. Intentionelle Schnittmarken auf Tierknochen von Bilzingsleben—Neue laser-mikroskopische Untersuchungen. *Praehistoria Thuringica* 3: 64–79.
Thévenin, A. 1976. Les civilisations du paléolithique intérieur en Alsace. In H. de Lumley (ed), *Le préhistoire française, Vol. 1: Les civilisations paléolithiques et m'esolithiques de la France*, pp. 984–996. Centre National de la Recherche Scientifique, Paris.
Thieullen, A. 1901. Os travaillés à l'époque de Chelles. *Bulletins de la Société d'anthropologie de Paris* 2(1): 347–362.
Valoch, K. 1987. The early Palaeolithic site Stránská skála I near Brno (Czechoslovakia). *Anthropologie* 25: 125–142.
Van Peer, P., R. Fullager, S. Stokes, R. M. Bailey, J. Moeyersons, F. Steenhoudt, A. Geerts, T. Vanderbeken, N. De Dapper and F. Geus 2003. The Early to Middle Stone Age transition and the emergence of modern behaviour at site 8-B-11, Sai Island, Sudan. *Journal of Human Evolution* 45(2): 187–193.
Vishnyatsky, L. B. 1994. 'Running ahead of time' in the development of Palaeolithic industries. *Antiquity* 68: 134-140.
Wason, P. C. 1960. On the failure to eliminate hypotheses in a conceptual task. *Quarterly Journal of Experimental Psychology* 12: 129–140.
Watson, B. 2009. Universal visions: neuroscience and recurrent characteristics of world palaeoart. PhD thesis, Centre for Classics and Archaeology, University of Melbourne.
Wendt, W. E. 1974. Art mobilier aus der Apollo 11 Grotte in Südwest-Afrika. *Acta Praehistorica et Archaeologica* 5: 1–42.
Ziegert, H. 1995. Das neue Bild des Urmenschen. *Uni hh forschung* 30: 9–15.
Ziegert, H. 2007. A new dawn for humanity: Lower Palaeolithic village life in Libya and Ethiopia. *Minerva* 18(4): 8–9.

Chapter 4
Seafaring, Beads and External Hard Drives

Heroes of the Dreamtime

One of the most sophisticated symbol systems developed by our lineage is of course language, and it is now widely agreed that maritime navigation and colonization of lands by seagoing vessels presupposes fairly complex communication forms, almost certainly of the verbal kind. Since Pleistocene seafaring necessarily involved forward planning and coordinated community efforts (Bednarik and Kuckenburg 1999), it is almost impossible to account for it in the absence of "reflective" language (Davidson and Noble 1989). Even the most dedicated SR (short-range) scholars have accepted this. After I pointed out to Professor Iain Davidson that his hypothesis of pictures of objects (appearing first, say, 32 ka ago) preceding the introduction of language (Davidson and Noble 1989) is contradicted by the earlier colonization of Australia (say, 50 or 60 ka ago), he changed his view and wrote a paper tellingly entitled "Why the first colonisation of the Australian region is the earliest evidence of modern human behaviour" (Davidson and Noble 1992). In accepting my argument that maritime colonization demands the availability of language he did, however, not realize that before people were able to reach Australia—from whichever island, as they first had to island-hop through the Indonesian archipelago—it involved several previous sea crossings. Moreover, he was unaware of the great time depth of seafaring practices: the first seafaring in the region occurred in the order of twenty times as long ago as he had thought. When it was pointed out (Bednarik 1995a) that hominins—*Homo erectus*, to be precise—had not only reached the island of Flores by the beginning of the Middle Pleistocene, but also established a thriving population there, a debate ensued that exposed some of the follies of the SR model. Davidson first sought to reject the dating of the Flores finds (a familiar pattern; he had previously sought to question the dating of the Bilzingsleben engravings, and when he realized his folly, he instead questioned their authenticity), and when he grasped his rashness he switched to a final-resort subterfuge and contended that the colonizers must have been washed out to sea involuntarily and had made all these crossings on floating vegetation mats.

He was not the only Australian archeologist who felt castigated by the revelation that the sea crossings of *Homo erectus* had been known for over 30 years, but had remained completely ignored in Australia. His colleague Professor Colin Groves

R.G. Bednarik, *The Human Condition*, Developments in Primatology: Progress and Prospects, DOI 10.1007/978-1-4419-9353-3_4,

chastised me severely for my impertinence of suggesting Australian archeologists were poorly informed but ended up conceding just that (Bednarik 1995a, b; Groves 1995). The problem had simply been that nearly all of the papers demonstrating the very early colonization of Flores had been published in German (Verhoeven 1958, 1964, 1968; Maringer and Verhoeven 1970a, b, 1972, 1975), and nothing appeared about this evidence and its implications in any English-language journal before the late 1990s (Bednarik 1997a, b, c, 1998a, b, 1999a, b, c, d, e, 2000, 2001a, b, c, 2002a, b, c, 2003, 2007a, b; Bednarik et al. 1999).

Seafaring in the Pleistocene has been demonstrated by several types of finds from about 20 islands that have never been connected to a mainland (most of them not even to another island), or at least not during the Pleistocene; and from the continent of Australia (Bednarik 2003). They consist of skeletal remains of approximately 200 humans, mostly from Australia but including those of nine individuals from four islands (Santa Rosa, Okinawa, Crete and Sardinia); and of human occupation evidence in the form of stone tools, food remains, ornaments, rock art, and campsites. The two main regions of Pleistocene maritime navigation evidence are the Mediterranean, where at least seven deep-water islands were occupied during the Ice Age, and the general region of eastern Asia to Australia. The only other island with known Pleistocene occupation is Santa Rosa, one of the Californian Channel Islands.

The earliest Mediterranean evidence, Clactonian stone tools found in Sardinia, is thought to be in the order of 300 ka old (Bini et al. 1993). One cave site has provided a single bone of *Homo erectus*, our evolutionary predecessor (Ginesu et al. 2003). The recent discovery of handaxes on Crete (Mortensen 2008; Kopaka and Matzanas 2009; Strasser et al. 2010, 2011) implies the presence of Lower Paleolithic traditions on that island and raises the possibility of African contact. However, by far the most extensive and the oldest proof of seafaring comes from Indonesia, where this technology was probably first developed. Evidence of hominin occupation of the late Lower Pleistocene and Middle Pleistocene is now available from Selatan (Sulawesi) (M. Morwood, personal communication, Dec. 2009) and three islands of Nusa Tenggara, formerly called the Lesser Sunda Islands: Flores (Verhoeven 1958; Maringer and Verhoeven 1970a; Sondaar et al. 1994), Roti (Bednarik 1999a), and Timor (Bednarik and Kuckenburg 1999). The early stone tools of Flores have been shown to be up to 840 ka old by a variety of dating methods, including fission-track dating, paleomagnetism, geology, paleontology, and the presence of datable tektites. Some of the strata containing these finds of stone tools, and in the case of Timor charred animal remains, are overlain by substantial facies of solid Quaternary rock, in places of more than 150 m thickness (Bednarik 2003). Their great age is, therefore, beyond dispute, irrespective of the comprehensive scientific dating evidence.

The presence of *Homo erectus* is thus amply demonstrated in Wallacea. It is the only human species known to have existed at that time. The islands of Wallacea, of the tens of thousands of islands between Asia and Australia/New Guinea, are named after Alfred Russel Wallace, the nineteenth century British naturalist. He discovered the almost complete absence of Asian mammals on these islands and

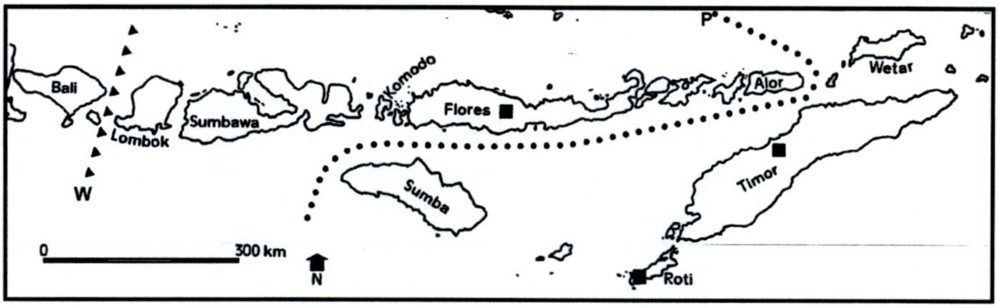

Fig. 4.1 Map of southern Wallacea (Nusa Tenggara), Indonesia. The presumed dividing line between the Eurasian and Australian continental plates is shown. Wallace's biogeographical line runs between Bali and Lombok, and the locations of the known hominin occupation sites of the Early and Middle Pleistocene are indicated by *squares*

correctly deduced that this fauna never had the opportunity of colonizing them (Wallace 1890). The Wallace Line is in fact the most important biogeographical barrier or filter in the world. It runs between Bali, which at the lower sea levels of the Pleistocene was connected to Asia via Java and Sumatra, and Lombok, the first of the islands that were never part of any other landmass. Indeed, all of these islands are geologically very young, having been formed as the Australian plate slid under the Asian plate in the late Tertiary. They are only a few million years old (Fig. 4.1).

To reach Flores, human colonizers had to first cross to Lombok, from there to Sumbawa, and thence via Komodo to Flores. Alternatively they could have arrived via Selatan. These crossings demanded the use of watercraft, particularly as each had to be completed by a genetically viable breeding population including an adequate number of fertile females. In all cases within Nusa Tenggara the opposite shores were clearly visible from the shore of departure—at any sea level of the Ice Age period.

The first journey to Australia, however, occurred significantly later, possibly about 60,000 years ago (Roberts et al. 1993). This great delay was probably due to the fact that Australia has never been visible from Timor or Roti. The north of this continent is comparatively flat, and especially the very wide continental shelf facing Timor, which was exposed during periods of low sea level, had no elevations of much more than 100 m height (Bednarik and Kuckenburg 1999). Consequently the shore became only visible after about nine-tenths of the distance from Timor had been crossed. This does not mean, however, that the Pleistocene inhabitants of Timor could not know the presence of a large landmass to their south. They would have been experienced mariners, attuned to recognizing indirect evidence of land. For instance, smoke of large bushfires, which may rise several kilometers into the air, would have been visible occasionally, and in the tropics typical cumulus cloud formations often develop over islands, especially during afternoons. In addition, Pleistocene mariners would have observed the directions of sea currents, prevailing wave direction, and the seasonal or migratory movements of birds and sea creatures, such as tortoises. The crossing to Australia was, therefore, probably delayed for many hundreds of millennia because it demanded not only a developed symbolic tradition capable of placing sufficient trust in the conscious judgment of signs to accept the risks involved in sailing to an unseen destination, but it also required immense personal courage. These mariners were truly Heroes of the Dreamtime.

The first seafarers can thus be assumed to have been enticed by visual contact with the target shore, and this applied to most of the known crossings that occurred prior to the first landfall of Australia. The actual distances to be journeyed in each case are not readily known, because the time the crossings first occurred is not precisely established, nor are the exact sea levels at these times. Moreover, the elasticity of the Earth's crust and the processes of isostatic compensation is a variable which it is difficult to account for with precision, and the high incidence of tectonic adjustments in the general region of Indonesia throughout the late Tertiary and the entire Quaternary with its attendant frequent vulcanism would resist any simplistic modeling. It is, therefore, much safer to speculate about sea distances by determining which large land eutherians had managed to breach the sea barrier in question. By far the best maritime colonizers are elephants (capable of swimming 50 km on the open sea, and probably more), followed by hippos (20 km) (Bednarik and Kuckenburg 1999). Deer, bears, tapirs, rhinos, and pigs, although excellent swimmers, only manage considerably shorter sea distances of 5–15 km (particularly under adverse maritime conditions), even though they and other genera taking to the water readily all occurred in Pleistocene Southeast Asia. Small species (rodents, lizards, snakes, etc.) are not relevant here, their ability to raft on floating vegetation for long periods is well known (Diamond 1977).

Various species of proboscideans crossed to numerous of the islands of Wallacea and the Philippines where they experienced speciation and dwarfism in many separate developments (Hooijer 1957; Verhoeven 1964; Groves 1976; Bednarik 1999a; Bednarik and Kuckenburg 1999). Elephants are superb long-distance swimmers, having been observed to swim for 48 h in herd formation across African lakes, and in one reported case swimming a distance of 48 km at sea and in another at a speed of 2.7 km/h (Johnson 1980). Their buoyancy is helped by digestive gases in their intestines and their habit of traveling as a herd would facilitate the success of a founding population upon landfall. On this basis it is a reasonable assumption that the distance between Bali and Lombok, or anywhere along the Wallace Line, was never much less than 30 km, i.e., similar to that of today. Humans were the only other large mammals to cross these sea barriers, but they do not cross water barriers in herd formation by swimming, nor do they possess trunks to rise above the water or digestive gases to assist buoyancy; they are in fact poor swimmers by nature. If it were possible to cross sea straits on mats of floating vegetation, as Davidson has contended, other large species would have done the same, so this is not a viable interpretation either (if he were serious about this capricious claim he would demonstrate that it is possible to cross sea narrows by drifting on a vegetation mat). Consequently there is only one possible explanation of the evidence: humans developed the ability to construct watercraft very significantly earlier than has been considered possible.

This explanation is supported by the numerous subsequent Pleistocene colonizations of islands, by that of Australia, and probably even by the first entry into

Europe, which on the basis of current evidence could well have been via the Strait of Gibraltar rather than from Asia (Bednarik 1999d, 2001a).

In much the same fashion as we can use the absence or presence of certain species on deep-water islands to formulate tentative propositions about the former distances of sea to be crossed to reach them, we can also use the demonstrated capability of humans to cross sea barriers to estimate their technological capabilities at a given point in time. This approach of determining levels of technological competence is revolutionary in archeology, and it promises much better resolution than the traditional approach of basing such assumptions mostly on the contents of refuse deposits and discards. In the same way as the maximal capabilities of modern society, such as space travel, are unlikely to be reflected by the debris found in our garbage dumps, this is a much more precise way of fathoming the past.

The First Mariners Project (Bednarik 2003) has made its task precisely to determine the minimum preconditions under which colonizing parties would have been capable of crossing specific sea barriers. Since we know roughly at what time these crossings were first accomplished, this should then tell us what the people concerned were capable of—not only in a technological sense (resources, tools, techniques), but also in a cognitive and cultural sense. The skills necessary for such feats cannot be passed on genetically; they need to be conveyed culturally (culture referring here to the non-genetic transfer of practice). To do so, certain prerequisites are essential in the case of maritime navigation skills. To determine what these were is an integral part of cognitive archeology.

The First Mariners Project

Obviously this kind of study demands extensive replication experiments. These need to include the design and construction of the most primitive sea-going vessels by which the crossing in question is achievable. Such experiments cannot produce true replicas, but by using in the construction of such watercraft the materials and tools available at the time and in the region, the range of possibilities is narrowed down very considerably. Moreover, the rationale of the experiments is not so much to establish how a crossing can succeed, but to determine the practical limits of how it would be possible, by instead finding out at what technological level it would fail. This approach mirrors the principle of falsification, the basis of modern science, and for this reason alone it is a scientific procedure.

Such a program of replication must also include a great many further experiments. For instance, it needs to be shown how drinking water can be carried on sea journeys, and on longer trips how the sailors can carry or obtain sustenance. One would expect them to be accomplished fishermen; otherwise they would presumably not have developed maritime skills in the first place. To replicate the way they would have fed themselves at sea one needs to establish how to make harpoon points of types known to have been used at certain times in human history, using the tools then available. This is only one example and many other such replicative experiments copying artifacts of the time have assisted the First Mariners Project greatly in securing an understanding of available technologies. These need to cover all aspects of procuring, curating, treating, storing, working, and using the materials presumably in use at the time. Such materials might include bamboo, wood, bark, vines, fibers, wax, resins, bone, stone, skins, and leaves or palm fronds.

This renders the experiments underpinning the research into earliest seafaring very complex indeed. The principal tools to be used in all these related activities—from preparing the tools for fire-making by friction (Fig. 4.2) to setting tools in resins, to splicing cordage of split forest vines and making wooden paddles—are stone artifacts of the period in question. For crossing the sea barriers within Indonesia, to Sardinia or the Strait of Gibraltar, the tool models have to be stone artifacts of the Lower Paleolithic. For the crossings to Australia, Okinawa, and several other Mediterranean islands, Middle Paleolithic or Mode 3 tools would have been used, and replicas of them must be the basis of any such experimentation.

Based on these rationales, the perhaps largest project ever undertaken in replicative archeology was commenced in 1996. It consists of numerous experiments conducted under controlled conditions, from the production of stone tools to the construction of many simple but seagoing rafts. An integral part of this extensive program is to study the results and by-products of many work processes, such as microscopic wear traces on stone tools used exclusively for specific tasks. The highlights of the project, however, are the sea-trials of completed vessels, and the attempts in some cases to cross specific sea barriers.

So far, the First Mariners Project has constructed eight sea-going vessels, six in Indonesia and two in Morocco (Bednarik 1998b, 1999a, b, c, e, 2001b, 2003, 2007a; Bednarik and Kuckenburg 1999). Of these, two were models to test technologies; six were used to attempt full-scale crossings, of which four were completed successfully and two failed. The story of this venture is not just a scientific adventure in the finest

Fig. 4.2 Fire-making equipment used on board of replicative bamboo raft of the First Mariners Project

tradition; it has generated a knowledge base about Pleistocene seafaring that did not exist at all before. The use of sea-going rafts faded out about a hundred years ago in Indonesia, before detailed observations were made of the effectiveness of the relevant technology (Hornell 1946; Ling 1956, 1970; Doran 1971, 1978). It now has to be re-learnt, starting from its most basic aspects.

The first experimental raft, built with stone tools in late 1997 and early 1998, was the 15-ton, 23-m *Nale Tasih 1*, launched in Roti on 14 February 1998 (Bednarik 1998b). A number of design shortcomings and material defects led to the abandonment of the attempt to sail the raft with a crew of eleven to Australia. A radically different, simpler design was adopted for *Nale Tasih 2*, an 18-m bamboo raft of only 3.6 tons. Construction of this vessel began in August 1998 near Kupang, West Timor, and the primitive raft was launched in mid-November. On 17 December it left Kupang with a crew of five (Bednarik and Kuckenburg 1999). It was well equipped with spare parts, including sails made of palm leaves, a steering oar, forest vines for cordage, and to effect repairs it carried 65 stone artifacts. Several harpoons of MSA design were used to catch fish. The raft reached the continental shelf of Australia, which formed the continent's shore 60,000 years ago, after six days, thus having completed its primary objective (Fig. 4.3). To gain more knowledge in its handling, the crew continued on toward Darwin. On the eleventh day, seas became rough and the raft was sailed under extreme conditions for two days (Fig. 4.4). The steering oar broke; a yard broke in two, and at one stage, all forward guy ropes of the mast snapped in unison, which created a dangerous situation. However, all repairs were effected successfully. After 13 days, rough seas of 4-5-m waves forced the raft toward Melville Island, north of Darwin (Bednarik and Kuckenburg 1999).

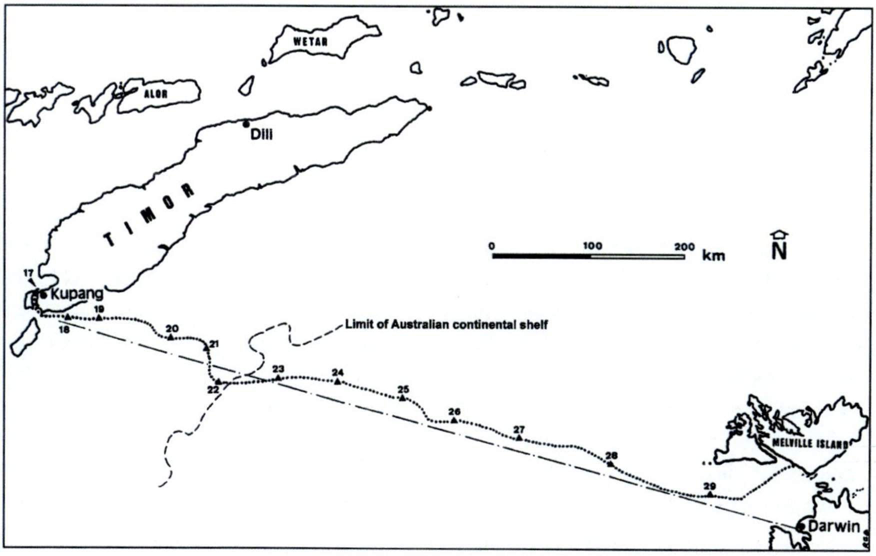

Fig. 4.3 Course of the *Nale Tasih 2* from Timor to Australia, in December 1998

Fig. 4.4 The *Nale Tasih 2* battling 5-m waves in tropical storm near Australia, 28 December 1998

The *Nale Tasih 3* was an 11.4-m-long bamboo raft propelled by six oars (Bednarik 2001b). On 23 March 1999, six oarsmen tried to row the simple platform from Bali to Lombok, over 35 km away. Every effort was made to row against strong transverse currents, but the attempt had to be abandoned under appalling conditions. A second attempt was made on 31 January 2000 (Bednarik 2001b). A simple bamboo platform with 12 paddlers just barely managed the crossing in 12 h. Most of the crew experienced episodes of severe fatigue, one of us collapsed unconsciously and remained in a coma for two days (Fig. 4.5).

Two further rafts were then built in Morocco, on a beach facing Spain across the Strait of Gibraltar. One was a pontoon raft made from cane, the other consisted of inflated animal skins (Bednarik 2001a). Both were constructed entirely with stone tools of Lower Paleolithic types of the time when Europe was first settled by humans from Africa. Both these rafts were tested at sea but no attempt was made to cross to Spain.

A more recent experiment sought to establish the minimum material requirements of bringing ten people from Sumbawa to Komodo (the latter island was joined to Flores during a period of lower Pleistocene sea levels), using the means available to Lower Paleolithic hominins (Bednarik 2007a). This crossing of Sape Strait took place in October 2004. The *Rangki Papa* ("Father of Rafts") embodied all that had by then been learnt about seagoing rafts, and it performed superbly. Nevertheless, the powerful transverse currents again prevented reaching the western coast of the target land, despite calm and most favorable conditions. Notwithstanding all efforts to maintain an easterly course, the vessel drifted south to the open sea and could reach Komodo only by sheltering from the current below the southwestern corner of that island, on the lee of the ramparts of Pula Lankoi (Fig. 4.6). This demonstrated again how impossible it must have been, at any time, to cross a narrow strait without

Fig. 4.5 Om Mberu on the *Nale Tasih 4*, Lombok Strait, 31 January 2000

Fig. 4.6 The *Rangki Papa* reaching the cliffs of Pula Lankoi, western Komodo, on 7 October 2004

an effective propellant. Such island hopping cannot be accomplished by mere drifting, be it on a vessel or on naturally accumulated vegetation matter—at least not in a reasonable time span. Such sea crossings were only possible for small animals, such as rodents, lizards, snakes, and insects, capable of living on drifts for periods

Fig. 4.7 The *Lombok* at the longitude of Pulau Dua, approaching the west coast of Sumbawa, 5 March 2008. The author is second from the *right*

of months and years, in the same way as many types of plant seeds are capable of surviving at sea. Such natural drifts can travel great distances, but as already Wallace noticed, they cannot support larger mammals, which explains their absence in Wallacea.

For the attempt to cross the remaining water barrier to Flores, Ales Strait, which separates Lombok from Sumbawa, the 13.6-m-long *Lombok* was built in 2008. It was the best performing raft of the project so far, the product of years of systematic studies (Fig. 4.7). In defiance of expert advice, it was constructed from green (uncured) bamboo stalks; indeed, throughout the First Mariners Project most academic expert advice had been found to be flawed, whereas that of local fishermen and traditional boat-builders was inevitably validated by the replication experiments. On 5 March 2008, a handpicked crew of eight set out from north of Labuhanhai, crossing the strait north of Pulau Dua in the record time of under 6 h. But with the beach only 300 m away, a maelstrom turned the vessel around and it raced at 20 or 30 knots back west, forced by an irresistible current. The crew tried desperately for the next 5 h to land further south, before succeeding at last. Once again, the treacherous nature of currents in narrows was demonstrated graphically. The superbly performing raft was then towed to Flores for further sea trials (Fig. 4.8).

This project is to continue for several more years but some distinctive trends have already emerged. These patterns suggest that the distances traveled by hominins increased gradually over time, as did the severity of other obstacles mastered

Fig. 4.8 The *Lombok* conducting trials and experiments on the south coast of Flores, 13 April 2008

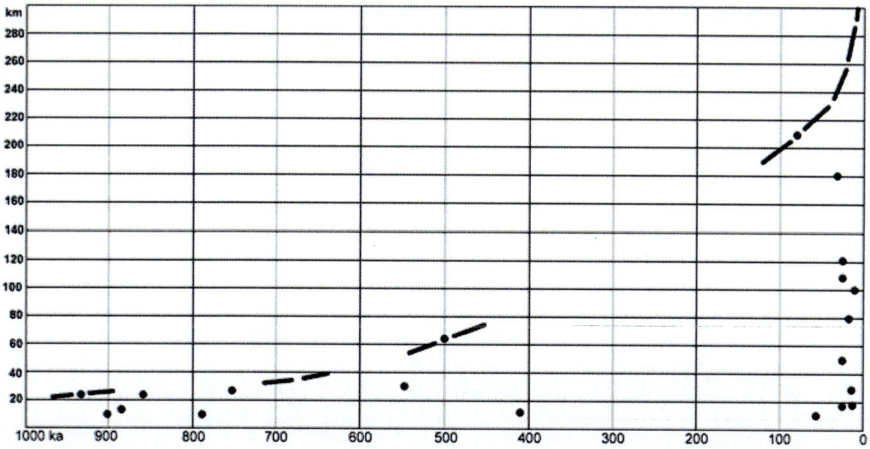

Fig. 4.9 Estimated time of first maritime colonizations (in ka) plotted against presumed shore distances at time of travel, showing how maximal distances traveled increased gradually through time. The *broken line* thus indicates approximate maximal navigation capability of humans through time

(Fig. 4.9). By about a million years ago we can assume that the first use of navigation was emerging in the area of Java and Bali, probably to exploit offshore fishing grounds. This led to the ability of crossing perhaps over 30 km of water with colonizing groups. By 840 ka BP, at least three such sea barriers had been breached and

Flores had an archeologically highly visible population of *Homo erectus*. Perhaps around the same time, if not earlier (the issue of the first colonization of Europe remains controversial), Acheulian mariners from northwestern Africa managed to cross to Gibraltar, a distance of only between 10 and 14 km at lower sea level, and possibly even reached Crete. The Indonesian sailors eventually conquered a distance of between 60 and 100 km, from Flores or Alor to Timor, still in the Middle Pleistocene.

Much more development was required before humans were ready to cross to the unseen landmass of Sahul, which involved an actual distance of at least 200 km (the direct distance to the continental shelf is not relevant, because crossing along the shortest route would have always been impossible). This occurred much more recently, probably in the order of 50–60 ka ago. But by 30 ka before the present, distances in excess of 100 km were crossed frequently, and often to quite small islands, even in both directions (Bednarik 2003). By that time, seafaring would have still involved very high risks, but the settlement of around 20 islands that we know of (and no doubt others where we have yet to look) suggests that maritime navigation had reached the level of taking calculated risks. While it was now well developed in two world regions and spreading to others, the most sophisticated tradition remained that of the area to the north of Australia, eventually extending further into the Pacific, and it probably remained the most developed maritime technology in the world until modern history.

In the context of probing the cognitive evolution of hominins their early maritime prowess permits insights into several incidental issues. Seafaring is the earliest example we have in human history of the domestication of multiple natural systems of energy. It harnesses the combined effects of four forces of nature—those of the waves, currents, wind, and buoyancy—and it remains the most complex utilization of energy systems throughout the Pleistocene period (Fig. 4.10). Until the inventions of wheel and sledge it also remained the only mode of assisted locomotion used on

Fig. 4.10 Maritime colonization provides the earliest clear indication of "human modernity"

this planet ("assisted" in contrast to autonomous locomotion, as in walking, running, crawling or swimming). It would have promoted the formation of new neural structures on a scale not seen hitherto, such as those supporting "conscious" awareness of cause-and-effect relationships. This, too, has neurobiological implications for symboling abilities.

Still other abilities seem to be evident from these developments. For instance, the need for forward planning implies that concepts of time were a shared social reality, probably reified in some communicable form. Other technologically suggested variables refer to the need for cordage, and thus for knotting, without which no form of simple watercraft (almost certainly types of bamboo rafts were involved initially) can effectively be constructed. Cordage is of course also necessary in the use of other, more direct indicators of symbolism: beads and pendants.

Bringing Beads to the Natives

Among the most important empirical sources concerning the cognitive evolution of hominins are beads and pendants. A cardinal error of the SR advocates is that it is assumed beads, like other forms of paleoart, are entirely limited to "anatomically modern humans": they were brought to the aborigines of Europe by the "Moderns" from Africa 40 ka ago. The importance of beads is that, firstly, their use demonstrates self-consciousness with all its implications, itself an important factor in cognitive evolution. But they also demand the existence and communication of complex symbolic meanings, without which beads are of no use. Whatever their practical purpose may have been (decorative, communicative, emblemic, economic, protective, commemorative, ideological, etc.), their function was always deeply symbolic; they demonstrate essentially modern cognition, a property totally separate from considerations of physical evolution or technology.

Fortunately the secure identification of beads and pendants is largely uncontroversial. One of the principal arguments leveled against evidence suggestive of very early symbolism is that there are perfectly valid alternative explanations. This is indeed often the case. Natural surface markings of portable objects of various types have been misinterpreted as meaningful engravings in literally thousands of cases worldwide. I have examined and rejected hundreds of instances (600 in China alone). By far the most common examples are objects of bone, limestone, ivory, and ostrich eggshell. For instance, I have shown that they often bear mycorrhizal grooves that may resemble engravings (Bednarik 1992). Bone fragments can feature markings made by animal canines, by gastric acids (e.g., of hyenas), or by other taphonomic agents of various types (trampling, sediment movement, solifluction, cryoturbation, etc.). Other very common examples are perforated bone fragments and shells, which some archeologists have interpreted as anthropic products—intentionally made by humans. Bones can be perforated by animal teeth and corrosive agents, gastropod shells are commonly bored through by parasitic organisms. Similarly, natural surface markings on rock have often been archeologically misinterpreted, and again I have corrected more such instances than anyone, in which either natural markings were identified as rock art or rock art as natural markings (Bednarik 1994).

Some commentators on the issue of whether perforations of Pleistocene objects were natural or artificial apparently make a fundamental error of logic (d'Errico and Villa 1997). They seem to believe that, in order to be considered to have been used as a bead, a perforated object must have been *made* by humans. Any consideration of the kinds of objects used as ethnographic beads will readily show this to be false. The correct logic is that one may be able to demonstrate the use of a bead in some cases from microscopic evidence (Bednarik 1997d, 2005), but one can never demonstrate that any perforated small object found in an occupation layer was *not* used as a bead. In view of the widespread use of beads today, and the frequency with which they are lost, and considering further that beads were in use for some hundreds of millennia (as we have seen in the previous chapter), almost certainly in large quantities, it is very much more likely than not likely that most perforated small objects found in an occupation layer were used as beads. The fact that we cannot *prove* that a naturally perforated, bead-like object was used as a bead should not prompt us simply to exclude it from any consideration.

The outstanding characteristic of made beads and pendants is that their archeological identification is usually unambiguous, which one cannot always say about other classes of symboling evidence. Small objects, drilled through with stone tools, could be either beads or pendants, or they could be small utilitarian objects such as buckles or pulling handles, or the quangings (pulling handles used in sealing) of the Inuit (Boas 1888: figures 15, 17, 121d; Nelson 1899: Pl. 17; Kroeber 1900: figure 8). Such utilitarian objects are generally of distinctive shape, use wear, and material; they need to be very robust. Small objects that were drilled through either in the center or close to one end (e.g., teeth perforated near the root), that are too small or too fragile to be utilitarian objects, and that lack the typical wear patterns of such articles, can be safely assumed to be beads or pendants. The evidence that they were drilled with a stone tool is often indicated by a distinctive bi-conical and "machined" section and sometimes by rotation striae. The wear of pendants can often be observed on archeological specimens, including those made of stone (Bednarik 1997d), and is also quite typical.

An example of such complete lack of ambiguity are the disc beads made from ostrich eggshell. These are extremely common in the ethnography of southern African people (Woodhouse 1997), and in the archeological record they are found from there to China and Siberia (Bednarik 1993a). The ostrich (*Struthio camelus* ssp.), now extinct in Asia, was widespread in much of Africa and Asia to the end of the Pleistocene, even into the Holocene (at least in Arabia; Bednarik and Khan 2005). Its eggshell was used widely, as containers and as decorative material. In southern Africa, such use extends from the present back to the "Middle Stone Age" (Mode 3 technocomplex). Decorated fragments have been reported from the Howieson's Poort phase in Apollo 11 Cave, Namibia (Wendt 1974), from the MSA of Diepkloof Cave in the southwestern Cape area (Beaumont 1992) and as beads from Bushman Shelter in Transvaal (Woodhouse 1997), both in South Africa, and from White Paintings Rockshelter in the Tsodilo Hills, Botswana (Morris 2000); and two ostrich eggshell beads and several fragments, MSA but undated, presumed between 45 and 280 ka, are from Loiyangalani River Valley (Serengeti National Park, Tanzania). Some of these may be up to 80 ka old, and many more recent traditions have used such disc beads. Then there is a broken circular ostrich eggshell pendant, 3 cm in diameter and with a central perforation, from Bed 9 containing "Late Pietersburg," plus segments at the Cave of Hearths (Mason 1988), with an estimated age of 76 ka on the basis of dates for comparable Stratum 1 RGBS lithics at Border Cave (Grün and Beaumont 2001).

In Tunisia and Algeria, Capsian occupation deposits have yielded ostrich eggshell beads frequently, and these date from the very early Holocene. In India, deposits of the Late Pleistocene at 41 sites have produced ostrich eggshell fragments, and radiocarbon dates derived from such fragments range from 39 to 25 ka (Kumar et al. 1988). At two sites, Patne and Bhimbetka, a few disc beads have been found (Bednarik 1993a, b). The two specimens from Bhimbetka come from the neck region of a human burial, which suggests that they may have formed part of a necklace (Fig. 4.11). Similar beads occur in the Gobi desert, where they are found among the occupation remains of an Epipaleolithic or even Mesolithic tool tradition usually named after the site of Shabarak-usu (Bednarik and You 1991). Further finds of ostrich eggshell disc beads, of roughly similar age (final Pleistocene to early Holocene), have also been reported from Inner Mongolia (Hutouliang) and southern Siberia (Krasnyi Yar, Trans-Baykal).

Of a significantly greater antiquity are the over forty similar ostrich eggshell beads from El Greifa site E, in Wadi el Adjal, Libya (Bednarik 1997d; Ziegert 2007). As noted in the previous chapter, they are from Acheulian occupation deposits, which represent many millennia of continuous occupation of a littoral site, on the shore of the huge Fezzan Lake of the Pleistocene. This site has exceptionally good preservation conditions, with insect remains and seeds found together with bone. The typical Late Acheulian stone tool forms, including "handaxes," confirm the

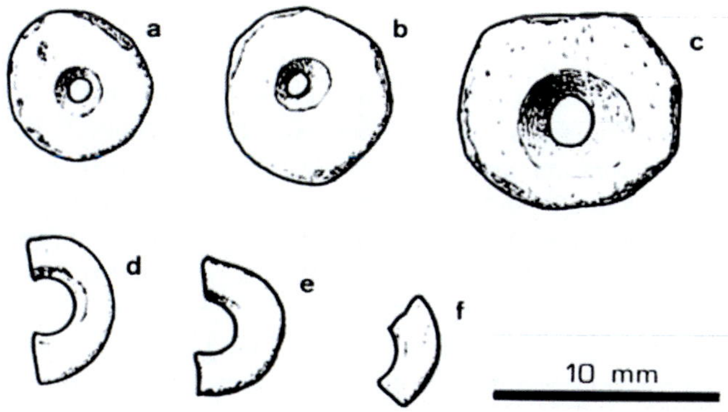

Fig. 4.11 Ostrich eggshell beads from the Upper Paleolithic of Bhimbetka (**a**, **b**) and Patne (**c**), India; and three from the Acheulian of El Greifa site E, Libya (**d–f**)

dating of the occupation strata by Th/U analysis to about 200 ka. These are the earliest known secure disc beads in the world, and there can be no reasonable doubt that they are indeed man-made beads and not some chance product of nature (Fig. 4.11).

Some SR advocates have questioned the dating of the El Greifa beads, which again illustrates the accommodative thinking of this school. The dating is, after all, only supplementary information; *any* Acheulian finds are unlikely to be significantly younger than 200 ka, and technologically they always relate to a Mode 2 tradition. Moreover, the Libyan beads may well be exceeded in age by many other finds, such as the hundreds of Acheulian stone beads mentioned in the previous chapter, or the two pendants from one of the occupation layers in the Repolust Cave, in the Austrian Alps. Here, however, the SR scholars (Rigaud et al. 2009) have opted for attempting to reject the anthropic agency. This means that the Styrian wolf incisor, perfectly perforated at its root (see Fig. 3.16), would be the only such tooth ever reported that was drilled by taphonomic agents. The question then arises: if this one specimen is the result of "partial digestion by carnivores," as d'Errico's team contends, how can we know that identically perforated teeth from younger deposits were not caused by the same factor? The answer, to the believers, is simple: if the finds are more than 40 ka old, the perforation must be natural, because people were then not capable of drilling through objects.

Interestingly enough, that does in their biased view not apply in Africa, at least not to material of the MSA, because it is in Africa that Eve's tribe evolved miraculously. Therefore, it is only natural to Eve's supporters that beads or pendants must have emerged in Africa before they were imported to Europe around 40 ka ago. There, the brutish Neanderthals of the Châtelperronian were so fascinated by them that they immediately began copying these purely symbolic artifacts, even scavenging them from the camps of the superior "Moderns" (e.g., White 1993; Hublin et al. 1996).

That is the gospel according to the Eve apostles: never let the facts get into the way of a good dogma. They fail to explain why a people incapable of symboling would find symbolic objects interesting, or what they would even do with them. This example shows not only how the believers can believe pretty well anything; it also illustrates how thoughtless this school of thought really is. Its adherents ignore that symbols have no meaning to other species; they seem unaware of what symbols are; why and how they denote propositions, terms, and relations in semiotic theory; or why and how they revolutionized human cognition and culture. The practice of deliberate body decoration, for instance, while extremely common in humans, is not observed in other animals who apparently have no comprehension of it.

Nevertheless, an incipient form of such behavior may have been observed by McGrew and Marchant (1998; McGrew 2004: 77, figure 5.1; Nishida et al. 2009) in 1996. A group of chimpanzees had killed and eaten a red colobus monkey one afternoon. The next morning, observers noted one of the juvenile females playing with a strip of colobus skin, and grooming it. It was then "stolen" from her by a juvenile male. The next morning a young adult female was observed wearing the skin strip draped around her neck. When she discarded it later that day the investigators found that the skin was tied in a single overhand knot, forming a simple necklace.

Perhaps the young adult female sought to enhance her appearance and/or status by adorning herself with the remains (symbol?) of a highly valued kill; or perhaps it was just a mistaken observation or an accidental occurrence. Be that as it may, the ability of non-human primates to tie simple knots is not new knowledge (Warner and Bednarik 1996), and numerous other embryonic forms of behavior traditionally thought to be limited to humans have been demonstrated in various species. Among them are toolmaking, forms of language-like communication, deliberate deception, self-awareness, domestication of other species, and the ability of interpreting others' intention. That the Neanderthals were brutish is entirely unsubstantiated, but it is empirically true that they had, on average, considerably larger brains than extant humans (up to between 1700 and 1800 ml). The relative size of an organ, we are told, is no reflection of its performance. But bearing in mind the enormous price humans pay for their oversized and energy-hungry brains, especially through their prolonged dependency enforced by their premature birth, it really does need to be explained what these brutes did with the contents of their skulls. Evolution would never permit the incredible enlargement of an organ unless there was a distinctive benefit attached to it. Perhaps SR scholars could elaborate.

Academic Tyranny

The SR advocates have made no attempt to explain the many paradoxes of their model. Indeed, the double standards Pleistocene archeology applies are so universal that they define the discipline. To illustrate: it may be true that there was no pre-Clovis human occupation of North America, but let us be quite clear that this has not been demonstrated. Irrespective of its incompatibility with South American evidence (Bednarik 1989), the argument is purely one based on the absence of evidence. As Charles Kingsley, the nineteenth-century English novelist, pointed out, "no one has a right to say that no water babies exist, till they have seen no water babies existing; which is quite a different thing, mind, from not seeing water babies". Yet the rule of absence of evidence not being evidence of absence is applied in a completely random fashion in Pleistocene archeology. For example, we have no hard evidence prior to 6000 BP that hominins possessed soft tissue or hair similar to our own, yet archeologists assume that they did; we have no material evidence of navigation prior to 9500 BP and yet it is generally accepted that Middle Paleolithic people used maritime vessels to reach Australia. In numerous cases, archeologists do not require hard evidence to accept the existence of a phenomenon or state; in others they require not only hard evidence, but demand large numbers of incidence (e.g., for symbolism); and even when that is furnished, they still reject it as inadequate. Not only does this indicate the exercise of unexplained double standards in the demands of evidence, the division between those phenomena requiring hard evidence as well as those not requiring it seems entirely random, in the sense that there seems to be no rationality in determining these categories. The universal rule appears to be: if evidence confirms the dogma, it must be valid; if it challenges the dogma, ways must be found to reject it, however much that may stretch the credibility of the discipline.

In the case of the rejection of symbolic evidence predating the "Aurignacian," the Humpty Dumptys of Pleistocene archeology, whose entirely etic terms (of tool types, cultures, traditions, peoples, ethnic groups, etc.) mean whatever they choose them to mean, have collectively fallen off the wall they had erected and sat on for far too long. All the king's horses and all the king's men cannot change that the entire replacement hypothesis, particularly the African Eve version, is nothing more than an academic sham. It is bereft of any real substance, was originally based on fake datings of fossils, was then transferred to unsupported genetic claims, sustained by accommodative hypotheses about invented and named tool industries and purported and named cultures, and was presented as a narrative rationalizing racism and genocide. But what is most disturbing about this incredibly naive notion is that the primary reason for its existence is simply archeological ignorance. It fulfils Albert Einstein's prediction that "the tyranny of the ignoramuses is insurmountable and assured for all time".

The endemic ignorance in the Anglo-American school of Pleistocene archeology is enough reason to be skeptical. A few years ago the English-speaking world marveled at these wonderful, "aerodynamically designed spears" from Schöningen in Germany, amazed how "Lower Paleolithic" artifacts could be so sophisticated. This was in response to their first publication in English, yet the Schöningen finds had long been published in German. Indeed, wooden spears of the "Lower Paleolithic" have been reported from seven sites in Europe, which also seems to have escaped the attention of these commentators. Moreover, it remains unknown to the Anglo-American school that an even more important find has been made at Schöningen, and apparently will remain so until it is published in *Nature* or *Science*. One of the favorites of the SR lobby are the portable engravings from Blombos Cave, which are rolled out every time the issue of early "art" comes up. This only shows that the commentators are ignorant about the much more richly decorated (on six of its seven facets) stone from Wonderwork Cave, which is of identical antiquity (Fig. 4.12).

Nearly all of the world's archeological data have never been published in any leading mainstream journal, and well over 80% of them have never been published in English at all. In epistemological parlance, language limitations are a major metamorphological factor in this field. There are hundreds of instances of relatively sound existing key archeological knowledge that I am aware of, but that are totally unknown in the ivory towers of Anglo-American academe—and yet they are utterly essential to forming a balanced picture of Pleistocene human history.

To any SR advocate who finds my critique unfair I offer this challenge: let us debate these issues openly and transparently. This should, firstly, be in public, and secondly, my opponents should not be permitted to comment on any piece of evidence they were not familiar with beforehand. The latter condition is to clarify that if they had no previous knowledge of the case, they forfeit the right to a biased rejection—which is their usual modus operandi. If I, a non-archeologist, can prevail in such debates, what does this tell us about the discipline?

Fig. 4.12 Engraved Middle Stone Age stone plaque from Wonderwerk Cave, South Africa, c. 70,000 years old

The replacement hypothesis is ultimately a cynical exercise of ethnic archeology, which has been driven out of more recent periods by political correctness (and the growing abilities of indigenes around the world to defend themselves), justifying the definition of some "other" as inferior and primitive, as the antithesis of our glorious ancestors. The African Eve model is simply the current archeological racist myth of how the world's benighted forces were defeated by those qualified to enter heaven. The coterie of its supporters would indignantly reject the notion that they might have been influenced by Christian values, but the thought-patterns of Europeans, and many others, bear the indelible imprint of that religion—however dedicated they may individually be as atheists or agnostics. We cannot escape the fact that our culture is largely based on Christian values (e.g., the calendar we use), which have in the past been the basis of slavery, fascism, pogroms, crusades, and inquisitions. We are individually barely aware how these dark forces impinge on our intellectual predisposition, yet we have the impertinence of judging other human societies and their values by our contingent biases.

At this point it has become apparent that the issues raised so far have some rather far-reaching implications, for Pleistocene archeology as well as the entire project of Western intellectuals judging societies they do not understand. Rather than continuing on this far too ambitious tangent, perhaps it is more appropriate in the present, limited context to return to the much simpler issue of taking to task the dominant model of how human modernity arose. A number of key questions have emerged, which the SR supporters, especially the Eve advocates, need to respond to:

1. What are the precise base-pair substitution rates in the human genome, the number of colonization events, and first colonization dates their genetic modeling is based on?
2. How do they propose to demonstrate that reticulate introgression did not take place, other than through the speculation that there were two separate species?

3. Where is their evidence that *contemporary* robust and gracile populations, anywhere in the world, were genetically so different that they could not produce fertile offspring?
4. How do they explain the clinal development from robust to gracile populations in four continents, and the common occurrence of intermediate forms?
5. Why do they reject their "genetic clock" when it is applied to another species?
6. How do they account for the preservation of "Neanderthal" autapomorphies in modern populations?
7. How do they explain their many false claims concerning various European human fossils, such as the Stetten specimens, Hahnöfersand, and Velika Pećina being of the Pleistocene?
8. Why did they fail to detect Protsch's fake datings for three decades?
9. How do they explain the presence of "Neanderthal" tracks in Chauvet Cave?
10. What is their evidence that the "Aurignacian" people were *fully* anatomically modern, bearing in mind that no people of the Pleistocene were so?
11. At what time did the purported invasion of Europe occur, based on precisely what archeological evidence?
12. How do they explain the complete absence of any evidence of expansion of a Mode 4 technology through northern Africa and the Levant until long after it was established in Eurasia?
13. Why are there no precursors to "Aurignacian" cave and mobiliary art along the purported trail of the invaders?
14. How do they account for the evidence that the early Mode 4 technologies of Eurasia developed in situ?
15. How do they explain that the invaders carried a Mode 4 technology to Eurasia, but a Mode 3 technology to Australia?
16. How do they account for the presence of fully gracile humans in Australia before they appeared in Europe?
17. How do they explain demographically that a physically stronger and more robust resident population covering all habitable regions of a rather frigid continent yielded to naked tropical invaders whose weaponry was identical, bringing with them a much smaller number of adaptive alleles?
18. If they were right that paleoart is limited to Mode 4 and later traditions, how do they explain the millions of occurrences of it with Mode 3 (and even earlier) industries?
19. How do they reconcile the claim that language is a late development with the evidence of numerous sea crossings by Mode 3 and earlier human groups?
20. What is the true status of the human fossils from Liang Bua in Flores? (If they cannot answer this question convincingly, a supplementary question would be: why should one take their other palaeoanthropological claims seriously?)
21. Why are they frequently unaware of evidence contradicting their ideas, and when it is pointed out to them, refuse to admit it to testing their hypothesis?
22. Why are they unaware that taphonomic logic not only accounts for the pattern observed in data recovery, it even demands such a pattern?

23. How do they account for the presence of "Neanderthal" genes in Europeans, Asians, and Papuans?

If the SR scholars cannot deal satisfactorily with these and many similar questions, and continue failing to address the significant shortcomings in their epistemology, is it not justified to see their model as the platform of an entrenched academic clique whose primary concern is to maintain its grip on academic hegemony in the discipline? The practitioners of a *scientific* discipline are expected to equitably consider contrary evidence, and while they may often fall short of their ideals, they do accept that a deliberate effort needs to be made toward impartiality—otherwise the entire scientific effort is hardly worthwhile. It would be disingenuous to pretend that such openness always governs academic discourse in the sciences, human nature being as flawed as it is (as this book will in the end demonstrate). Nevertheless, the intrinsic ideology of falsification does tend to cull unsound hypotheses, a force that is simply lacking in Pleistocene archeology, where the currency of hypotheses is determined by authority (*argumentum ad verecundiam*), historical fads, and social and political factors. When the enormous progress that has been made in scientific disciplines that are little more than half a century old, such as plate tectonics, ethology or applied genetics, is compared with the way Pleistocene archeology only seems to be chasing its own tail, the differences become starkly obvious. Kuhn (1962) reminds us that the term *science* is reserved for fields that do progress in obvious ways. Almost 180 years after the Pleistocene antiquity of humans was proposed (and noisily rejected by archeology for the next several decades), the experts still seem unable to clarify the simplest issues, such as the nature of some primate bones from a cave in Flores. They still argue whether Robusts had the language ability or cognition of an ape, or of a present-day human, or anything in between. They recycle the same tired old hypotheses, generation after generation, rejecting them, then refurbishing and reusing them, cycle after endless cycle—be it processualist, post-processualist, port-modernist or whatever. At the rate hominin species are being invented we will have hundreds of them in a few centuries, all with their impeccable but meaningless genomes and fanciful pedigrees. By comparison to the somatic differences between very disparate breeds of dogs (which are all of one species), the morphological differences between all of these humans will be negligible. While in theory Great Danes and Chihuahuas could produce fertile offspring not only with each other, but even with other so-called species, such as wolves or coyotes, we will have as many human species as once we had species of grizzly bears (300, and the grizzly is not even a separate species). Surely these many hominin "species" are as much taxonomic artifacts as are the archeofacts of stone tool typologies or Pleistocene "cultures": they are archeological fantasies. The African Eve fiasco is perhaps merely the tip of the iceberg, but it serves well in illustrating how archeology reacts to falsifying evidence. It does much the same today it did around one and a half centuries ago, when Boucher de Perthes complained:

> They employed against me a weapon more potent than objections, than criticism, than satire or even persecution—the weapon of disdain. They did not discuss my facts, they did not even take the trouble to deny them. They disregarded them.

Most of the paleoart evidence listed in the previous chapter has never been subjected to refutation attempts or even considered by the SR lobby. And when any of it has been examined, the objective was generally to preserve the dogma rather than begin with a null-hypothesis. This is absurd because the dogma can only be false, being clearly a reflection of highly fragmentary evidence derived from a non-random sampling process that is completely at the mercy of historical accidents in the sequence of knowledge acquisition. Hence, to defend a paradigm that can, by definition, only be premature, is entirely illogical.

To solicit the help of such an undisciplined discipline in determining how humans assumed those qualities that differentiate them from other primates is futile. It would seem unavoidable that in such a quest, the cultural dimensions of early hominin societies need to occupy a central position. Since Pleistocene archeology uses non-cultural variables to define its "cultures," it has little to offer in a serious pursuit of the question of how humans became human. Alternative avenues need to be explored, and I have chosen to focus on cultural productions and achievements. This has shown me that the dominant paradigm must be entirely false; that modern behavior could not possibly have appeared suddenly on the morning of the first Monday of the "Upper Paleolithic." It must have been in place for many hundreds of millennia, having developed over a very long time, and gradually. Hence, I am called a gradualist.

The Big Bang of Consciousness

Just 15 years ago that label "gradualist" amounted to a term of disparagement; it defined someone who had missed the train of enlightenment and, rather then conceding his backwardness, insisted on retarding the discipline's progress with his outdated notions. More recently, the cracks appearing in the fossilizing carapace of the replacement model have prompted some practitioners to review the state of the art and they have warmed markedly to the idea that the dogmatic Eve hypothesis is over-simplistic. This is not a development to be welcomed, however, because such a gradual change of perception permits a false dogma to fade away slowly, rather than implode in the paradigm shift that is required. This lingering can only lead to more contamination and to yet another round of debased hypotheses. It merely amounts to the reorientation of a "gestalt shift," or placing the same data "in a new system of relations with one another by giving them a different framework" (Kuhn 1962: 85), in lieu of the "paradigm shift" needed: "a noncumulative developmental episode in which an older paradigm is replaced in whole or in part by an incompatible new one" (Kuhn 1962: 92). Instead of Richard Klein's fictional big bang of consciousness of the final Pleistocene (Klein and Edgar 2002) we need a big bang of consciousness in twenty-first century Pleistocene archeology. Unfortunately, that is not likely to happen, and the discipline's inertia derives from the notion that slow retreat from a false model is better than a fresh, unencumbered start.

Curiously enough, the idea of an explosion in human development at an unspecified time between the "Middle Paleolithic" and the "Late Upper Paleolithic" seems self-contradictory. If it were attributable to the invasion of superior forces from Africa, why define it as a development? It would then merely manifest the practically instantaneous substitution of one suite of cultural characteristics with another, at a specific time. The evidence of such a big bang, if it was as profound as claimed, would surely be incisive, sudden, and striking; it would present some distinctive, easily datable proof. Yet instead of pinpointing such an event chronologically, the replacement advocates refer only vaguely to a period lasting up to 20,000 years, within which this big bang can be located anywhere according to one's preferred variables. Alternatively, one might contend that it occurred in Africa rather than Europe, but this would be at significant odds with the first occurrence of all the glorious achievements of Eve's progeny in Europe: the cave art, figurines, and Aurignacian technology have no precursors in Africa. Once again, we see that a key mistake of the Eve apostles was that they assumed modernity in behavior to have emerged in tandem with modernity in appearance—as they perceive it.

This raises the issue of defining hominin modernity in non-somatic terms: what is essentially "modern" human behavior? In this we need to bear in mind that archeology does not recover behavior, it merely invents interpretations of perceived behavior traces.

Symbol use and self-awareness are not defining variables of cognitive modernity, because both have been reported from other animals. Chimps will touch their foreheads when a dot of paint has been applied to it without their knowledge and they detect it in a mirror. This is taken to indicate their awareness of their own existence. Even elephants are thought to master this level, which is not surprising, given their extraordinary behavior in the face of the death of a family member. They may try to stand up a cadaver and their memory brings them back to handle the bones even after the body has decomposed (at this point is it useful to recall that some SR archeologists dispute the capacity of *Homo sapiens neanderthalensis* to "curate" their dead). We cannot yet know how widespread levels of self-awareness are in other species, but so far it has been attributed also to the remaining great apes, bottlenose dolphins, and magpies. The concept of "self-awareness" remains poorly defined in philosophy, psychology, and artificial intelligence, but it is taken to refer to an "explicit understanding of an individual that it exists." It may also include the understanding that other individuals are similarly self-aware, and it is absent in small infant humans. Much of what we tend to regard as self-awareness in extant humans may just relate to ideological, ontological, and cognitive a prioris, and to simple biological equipment, such as proprioceptors. The differences among species in levels of self-awareness may, therefore, not be as pronounced as we might be inclined to think. Similarly, the ability of various apes to master a range of symbols has been well established and, again, human superiority in symboling ability may not be as pronounced as we tend to think.

What perhaps facilitated the human ascent was the ability of humans to think a thought that no-one had ever thought before—a creative thought, which is only possible through the "advanced" (from our perspective) use of symbolic systems. The encephalization hominins experienced throughout their evolution is without question a most dramatic aspect of human development, resulting in the ultimate intellectual heavyweights, the "Neanderthals" and similar Robusts (e.g., the owner of the Narmada calotte): they developed the largest brains, relative to body size, of any creature on this planet. From <400 ml at 3–4 million years ago, to 800 ml some 1.7 Ma ago in *Homo erectus*, to over 1500 ml in *Homo sapiens neanderthalensis*, there is a relentless increase in brain size, the most distinctive enlargement occurring about 2 Ma ago in eastern Africa (especially apparent in cranial specimen KNM-ER 1470). Any theory of other developments of humans that does not factor in this incredible enlargement has no biological credibility. Most of this increase should be attributed to the evolutionary benefits bestowed upon progressively more skilled use of symbols: individuals capable of better communication and symboling, in whatever form, had more offspring. But biology imposed a limit on encephalization, and intelligence is a "last resort solution" because hard-wired behavior is energetically cheaper, so adaptive behavior would have benefits outstripping its neural cost.

Bearing in mind the enormous cost of encephalization to mothers and whole societies, it is illogical to maintain that these large brains were not used. First, the brain is by far the most energy-hungry organ in a human, consuming roughly a fifth of our energy without even having any moving parts. Secondly, in an evolutionary context it is simply inconceivable that an organism should have experienced such a dramatic change without adequate justification. The greatest disadvantages of the enlarging brain are the obstetric consequences at birth and the subsequent, significantly prolonged period of infancy. Duration of pregnancy, in any mammalian species, is a compromise between the demands of the need to expel the young as developed as possible, versus the vulnerability of the mother and the physiological limitation placed on the size of the birth canal. The larger the young is to be at birth, the greater the necessary pelvic space, which affects locomotion of the mother adversely, increases her dependency on the support of others, and further factors still, such as prolongation of breeding cycle. Since the fetus is essentially a parasitic organism that severely stresses the mother's body resources through its relentless demands, a longer pregnancy not only reduces the number of likely conceptions during her fertile life period, it also prolongs these physiological strains on the body of the mother. Alternatively, any shortening of the pregnancy would greatly prolong the dependency of the infant after birth, which may not only impair further breeding, but place additional demands on both the mother and the group. Both these factors impact greatly on evolutionary fitness: a reduction of the number of pregnancies over the fertile period of the female reduces overall reproductive success of the species, and the additional demands on the breeding group made by longer periods of weaning and other dependency (e.g., the need to be carried) further reduces the group's effectiveness. The most significant factor in this is brain size, the brain, and thus the cranium being relatively larger than other body parts at birth. Therefore, the price paid in evolutionary terms for the cortical enlargement in hominins was very substantial indeed.

The demand for further encephalization in humans could be circumvented by storing information, specifically of symbolic qualities, outside the brain, in a more reliably stable and relatively permanent form (Gregory 1970: 148). This idea of an external, surrogate "cortex"—or an external hard drive, as we might say today—was developed further by Goody (1977), Carruthers (1990), and Donald (1991: 124–161), but can be traced back all the way to Plato:

> If men learn this [writing], it will implant forgetfulness in their souls: they will cease to exercise memory because they will rely on that which is written, calling things to remembrance no longer from within themselves but from external marks (*Phaedrus*, 274e–275a).

Donald proposed a model positing three basic stages of human evolution. The first, according to him, is mimetic symbol use without symbol creation; the second is construction of conceptual space using language. The third involves the deposition of symbolic properties in material culture, capable of intervening in social behavior, or of communicating meaning. Of the obvious contenders of Donald's third type, style in tools and structured use of space have been proposed, but these may be illusory or only reflect the taxonomizing activities of archeologists. The major direct *surviving* evidence for such external storage of symbolic meaning is paleoart.

Donald's model, although it has given rise to a whole cottage industry called "extended mind (EM) studies" (Clark 2001, 2005; Clark and Chalmers 1998; Tollefsen 2006), has significant shortcomings (see e.g., debate in Donald et al. 1993; Donald 2001), and particularly lacks cohesively articulated material evidence (see Chapter 6). However, the concept of *external storage* is a goalpost of immediate value, even if it might seem to relegate the mind to the status of a Random Access Memory (Carruthers 1990: 7, 1998: 16; cf. Donald 1991: 310). If we accept such storage as the defining characteristic of cognitive modernity, it becomes fundamental in the context of the present considerations to establish how far such behavior extends into the past, and how it might have been acquired. It is perhaps this specific variable of cultural information stored in objects external to the brain, exograms, being symbolic which implies essentially "modern" cognitive faculties.

This brings the evidence we have visited in the previous chapter into a renewed, sharper focus. If we are to derive any empirically based answers to the question of modern human origins, this is the principal place to look for them: cognitive evolution, the beginnings of paleoart, the origins of symboling, and the external hard drive of cultural meanings are intertwined and collectively implicated in the formation of constructs of reality—to us the ultimate expressions of "conscious awareness." It is in this arena that humans have finally distanced themselves so far from all other animals that they can now modify their physical environment and harness its resources and energies on such a scale that they can determine the destiny of their planet. Humans not only learnt to construct or determine their physical environment, they also managed to create and navigate social structures and their space in ways other animals have not mastered.

Here I would like to examine how this initially came about and how we became human.

References

Beaumont, P. 1992. The time depth of aesthetic and symbolic behaviour in southern Africa. Paper presented to the Biennial Conference of the Southern African Association of Archaeologists, Johannesburg.

Bednarik, R. G. 1989. On the Pleistocene settlement of South America. *Antiquity* 63: 101–111.
Bednarik, R. G. 1992. Natural line markings on Palaeolithic objects. *Anthropologie* 30(3): 233–240.
Bednarik, R. G. 1993a. About Palaeolithic ostrich eggshell in India. *Indo-Pacific Prehistory Association Bulletin* 13: 34–43.
Bednarik, R. G. 1993b. Palaeolithic art in India. *Man and Environment* 18(2): 33–40.
Bednarik, R. G. 1994. The discrimination of rock markings. *Rock Art Research* 11: 23–44.
Bednarik, R. G. 1995a. Wallace's barrier and the language barrier in archaeology. *Bulletin of the Archaeological and Anthropological Society of Victoria* 1995(3): 6–9.
Bednarik, R. G. 1995b. Response to Colin Groves. *Bulletin of the Archaeological and Anthropological Society of Victoria* 1995(5): 8–11.
Bednarik, R. G. 1997a. The initial peopling of Wallacea and Sahul. *Anthropos* 92: 355–367.
Bednarik, R. G. 1997b. The origins of navigation and language. *The Artefact* 20: 16–56.
Bednarik, R. G. 1997c. The earliest evidence of ocean navigation. *The International Journal of Nautical Archaeology* 26(3): 183–191.
Bednarik, R. G. 1997d. The role of Pleistocene beads in documenting hominid cognition. *Rock Art Research* 14: 27–41.
Bednarik, R. G. 1998a. Mariners of the Pleistocene. *Institute of Nautical Archaeology Quarterly* 27(3): 7–15.
Bednarik, R. G. 1998b. An experiment in Pleistocene seafaring. *The International Journal of Nautical Archaeology* 27(2): 139–149.
Bednarik, R. G. 1999a. Maritime navigation in the Lower and Middle Palaeolithic. *Comptes Rendus de l'Académie des Sciences Paris, Earth and Planetary Sciences* 328: 559–563.
Bednarik, R. G. 1999b. Der Beginn der Seefahrt. *Almogaren* 30: 13–34.
Bednarik, R. G. 1999c. Sailing a Paleolithic raft. *Institute of Nautical Archaeology Quarterly* 28(1): 12–18.
Bednarik, R. G. 1999d. Pleistocene seafaring in the Mediterranean. *Anthropologie* 37(3): 275–282.
Bednarik, R. G. 1999e. Seefahrt im Pleistozän. *Quartär* 49/50: 95–109.
Bednarik, R. G. 2000. Crossing the Timor Sea by Middle Palaeolithic raft. *Anthropos* 95: 37–47.
Bednarik, R. G. 2001a. The origins of Pleistocene navigation in the Mediterranean: initial replicative experimentation. *Journal of Iberian Archaeology* 3: 11–23.
Bednarik, R. G. 2001b. Replicating the first known sea travel by humans: the Lower Pleistocene crossing of Lombok Strait. *Human Evolution* 16(3–4): 229–242.
Bednarik, R. G. 2001c. The dynamics of the first peopling of Wallacea and Australia. In J. Gillespie, S. Tupakka and C. deMille (eds), *On being first: cultural innovation and environmental consequences of first peopling*, pp. 81–91. Proceedings of the 31st Annual Chacmool Conference, University of Calgary, Calgary.
Bednarik, R. G. 2002a. The maritime dispersal of Pleistocene humans. *Migration and Diffusion* 3(10): 6–33.
Bednarik, R. G. 2002b. The first mariners. *The American Neptune* 61(3): 317–324.
Bednarik, R. G. 2002c. The First Mariners Project. *The Bulletin of the Australasian Institute for Maritime Archaeology* 26: 57–64.
Bednarik, R. G. 2003. Seafaring in the Pleistocene. *Cambridge Archaeological Journal* 13(1): 41–66.
Bednarik, R. G. 2005. Middle Pleistocene beads and symbolism. *Anthropos* 100(2): 537–552.
Bednarik, R. G. 2007a. Experimental crossing from Sumbawa to Komodo by bamboo raft. *INA Quarterly* 34(2): 13–17.
Bednarik, R. G. 2007b. Early seafaring. In H. Selin (ed), *Encyclopaedia of the history of science, technology, and medicine in non-Western cultures*, pp. 395–399. Springer Netherlands, Dordrecht.
Bednarik, R. G., B. Hobman and P. Rogers 1999. *Nale Tasih 2*: journey of a Middle Palaeolithic raft. *International Journal of Nautical Archaeology* 27(1): 25–33.

Bednarik, R. G. and M. Khan 2005. Scientific studies of Saudi Arabian rock art. *Rock Art Research* 22: 49–81.

Bednarik, R. G. and M. Kuckenburg 1999. *Nale Tasih: Eine Floßfahrt in die Steinzeit.* Jan Thorbecke Verlag, Stuttgart.

Bednarik, R. G. and Y. You 1991. Palaeolithic art from China. *Rock Art Research* 8: 119–123.

Bini, C., F. Martini, G. Pitzalis and A. Ulzega 1993. Sa Coa de Sa Multa e Sa Pedrosa Pantallinu: due 'Paleosuperfici' clactoniane in Sardegna. *Atti della XXX Riunione Scientifica, 'Paleosuperfici del Pleistocene e del primo Olicene in Italia, Processi si Formazione e Interpretazione', Venosa ed Isernia, 26–29 ottobre 1991*, pp. 179–197. Istituto Italiano di Preistoria e Protostoria, Firenze.

Boas, F. 1888. *The central Eskimo*, reprinted 1964. University of Nebraska Press, Lincoln.

Carruthers, M. 1990. *The book of memory.* Cambridge University Press, Cambridge.

Carruthers, M. 1998. *The craft of thought.* Cambridge University Press, Cambridge.

Clark, A. 2001. Reasons, robots, and the extended mind. *Mind and Language* 16: 121–145.

Clark, A. 2005. Intrinsic content, active memory and the extended mind. *Analysis* 65: 1–11.

Clark, A. and D. Chalmers 1998. 'The extended mind'. *Analysis* 58: 7–19.

Davidson, I. and W. Noble 1989. The archaeology of perception: traces of depiction and language. *Current Anthropology* 30(2): 125–155.

Davidson, I. and W. Noble 1992. Why the first colonisation of the Australian region is the earliest evidence of modern human behaviour. *Archaeology in Oceania* 27: 113–119.

d'Errico, F. and P. Villa 1997. Holes and grooves: the contribution of microscopy and taphonomy to the problem of art origins. *Journal of Human Evolution* 33: 1–31.

Diamond, J. M. 1977. Distributional strategies. In J. Allen, J. Golson and R. Jones (eds), *Sunda and Sahul: prehistoric studies in South-East Asia, Melanesia and Australia*, pp. 295–316. Academic, London.

Donald, M. 1991. *Origins of the modern mind: three stages in the evolution of culture and cognition.* Harvard University Press, Cambridge, MA.

Donald, M. 2001. *A mind so rare: the evolution of human consciousness.* W.W. Norton, New York, NY.

Donald, M. et al. 1993. Précis of 'Origins of the modern mind: three stages in the evolution of culture and cognition'. *Behavioral and Brain Sciences* 16: 737–791.

Doran, E. 1971. The sailing raft as a great tradition. In C. L. Riley, J. C. Kelley, C. W. Pennington and R. L. Rands (eds), *Man across the sea: problems of pre-Columbian contacts*, pp. 115–138. University of Texas Press, Austin, TX.

Doran, E. 1978. Seaworthiness of sailing rafts. *Anthropological Journal of Canada* 16(3): 17–22.

Ginesu, S., S. Sias and J. M. Cordy 2003. Morphological evolution of the Nurighe Cave (Logudoro, northern Sardinia, Italy) and the presence of man: first results. *Geografica Fisica e Dinamica Quaternaria* 26: 41–48.

Goody, J. 1977. *The domestication of the savage mind.* Cambridge University Press, Cambridge.

Gregory, R. L. 1970. *The intelligent eye.* Weidenfeld and Nicolson, London.

Groves, C. P. 1976. The origin of the mammalian fauna of Sulawesi (Celebes). *Zeitschrift für Säugetierkunde* 41: 201–216.

Groves, C. P. 1995. The origin of language, the use of language and other people's language. A response to Robert Bednarik. *Bulletin of the Archaeological and Anthropological Society of Victoria* 1995(4): 8–12.

Grün, R. and P. Beaumont 2001. Border Cave revisited: a revised ESR chronology. *Journal of Human Evolution* 40: 467–482.

Hooijer, D. A. 1957. A Stegodon from Flores. *Treubia* 24: 119–129.

Hornell, J. 1946. *Water transport: origins and early evolution.* Cambridge University Press, Cambridge.

Hublin, J.-J., F. Spoor, M. Braun, F. Zonneveld and S. Condemi 1996. A late Neanderthal associated with Upper Palaeolithic artefacts. *Nature* 381: 224–226.

Johnson, D. L. 1980. Problems in the land vertebrate zoogeography of certain islands and the swimming powers of elephants. *Journal of Biogeography* 7: 383–398.

Klein, R. G. and B. Edgar 2002. *The dawn of human culture: a bold new theory on what sparked the 'big bang' of human consciousness*. Wiley, New York, NY.

Kopaka, K. and C. Matzanas 2009. Palaeolithic industries from the island of Gavdos, near neighbour to Crete in Greece. *Antiquity* 83(321): http://antiquity.ac.uk/projgall/kopaka321/.

Kroeber, A. L. 1900. The Eskimo of Smith Sound. *Bulletin of the American Museum of Natural History* 12: 265–327.

Kuhn, T. 1962. *The structure of scientific revolutions*. University of Chicago Press, Chicago, IL.

Kumar, G., G. Narvare and R. K. Pancholi 1988. Engraved ostrich eggshell objects: new evidence of Upper Palaeolithic art in India. *Rock Art Research* 5: 43–53.

Ling, S.-S. 1956. Formosan sea-going raft and its origin in ancient China. *Bulletin, Institute of Ethnology, Academia Sinica* 1: 1–54.

Ling, S.-S. 1970. *A study of the raft, outrigger, double and deck canoes of ancient China, the Pacific and the Indian Oceans*. Monograph 16, Institute of Ethnology, Academia Sinica, Nanking, Taipei.

Maringer, J. and T. Verhoeven 1970a. Die Steinartefakte aus der Stegodon-Fossilschicht von Mengeruda auf Flores, Indonesien. *Anthropos* 65: 229–247.

Maringer, J. and T. Verhoeven 1970b. Die Oberflächenfunde aus dem Fossilgebiet von Mengeruda und Olabula auf Flores, Indonesien. *Anthropos* 65: 530–546.

Maringer, J. and T. Verhoeven 1972. Steingeräte aus dem Waiklau-Trockenbett bei Maumere auf Flores, Indonesien. Eine Patjitanian-artige Industrie auf der Insel Flores. *Anthropos* 67: 129–137.

Maringer, J. and T. Verhoeven 1975. Die Oberflächenfunde von Marokoak auf Flores, Indonesien. Ein weiterer altpaläolithischer Fundkomplex von Flores. *Anthropos* 70: 97–104.

Mason, R. 1988. Cave of Hearths, Makapansgat, Transvaal. *University of the Witwatersrand, Archaeological Research Unit, Occasional Papers* No. 21, 1–713.

McGrew, W. C. 2004. *The cultured chimpanzee*. Cambridge University Press, Cambridge.

McGrew, W. C. and L. F. Marchant 1998. Chimpanzee wears a knotted skin 'necklace'. *Pan African News* 5(1): 8–9.

Morris, D. 2000. Tsodilo—worked ostrich eggshell dated. *The Digging Stick* 17(2): 10.

Mortensen, P. 2008. Lower to Middle Palaeolithic artefacts from Loutró on the south coast of Crete. *Antiquity* 82(317): 1–6.

Nelson, E. W. 1899. The Eskimo about Bering Strait. In J. W. Powell (ed), *Eighteenth Annual Report of the Bureau of American Ethnology*. Government Printing Office, Washington, DC.

Nishida, T., T. Matsusaka and W. C. McGrew 2009. Emergence, propagation or disappearance of novel behavioral patterns in the habituated chimpanzees of Mahale: a review. *Primates* 50(1): 23–36.

Rigaud, S., F. d'Errico, M. Vanhaeren and C. Neumann 2009. Critical reassessment of putative Acheulean [sic] *Prosphaere globularis* beads. *Journal of Archaeological Science* 36: 25–34.

Roberts, R. G., R. Jones and M. A. Smith 1993. Optical dating at Deaf Adder Gorge, Northern Territory, indicates human occupation between 53,000 and 60,000 years ago. *Australian Archaeology* 37: 58–59.

Sondaar, P. Y., G. D. van den Bergh, B. Mubroto, F. Aziz, J. de Vos and U. L. Batu 1994. Middle Pleistocene faunal turnover and colonization of Flores (Indonesia) by *Homo erectus*. *Comptes Rendus de l'Académie des Sciences Paris* 319: 1255–1262.

Strasser, T. F., E. Panagopoulou, C. N. Runnels, P. M. Murray, N. Thompson, P. Karkanas, F. W. McCoy and K. W. Wegmann 2010. Stone age seafaring in the Mediterranean: evidence from the Plakias region for Lower Palaeolithic and Mesolithic Habitation of Crete. *Hesperia* 79: 145–190.

Strasser, T. F., C. Runnels, K. Wegmann, E. Panagopoulou, F. McCoy, C. Digregorio, P. Karkanas and N. Thompson 2011. Dating Palaeolithic sites in southwestern Crete, Greece. *Journal of Quaternary Science*. DOI: 10.1002/jqs.1482.

Tollefsen, D. P. 2006. From extended mind to collective mind. *Cognitive Systems Research* 7(2–3): 140–150.

Verhoeven, T. 1958. Pleistozäne Funde in Flores. *Anthropos* 53: 264–265.

Verhoeven, T. 1964. Stegodon-Fossilien auf der Insel Timor. *Anthropos* 59: 634.

Verhoeven, T. 1968. Vorgeschichtliche Forschungen auf Flores, Timor und Sumba. In *Anthropica: Gedenkschrift zum 100. Geburtstag von P. W. Schmidt*, pp. 393–403. Studia Instituti Anthropos No. 21, St. Augustin.

Wallace, A. R. 1890. *The Malay Archipelago*. Macmillan, London.

Warner, C. and R. G. Bednarik 1996. Pleistocene knotting. In J. C. Turner and P. van de Griend (eds), *History and science of knots*, pp. 3–18. World Scientific, Singapore.

Wendt, W. E. 1974. Art mobilier aus der Apollo 11 Grotte in Südwest-Afrika. *Acta Praehistorica et Archaeologica* 5: 1–42.

White, R. 1993. Technological and social dimensions of Aurignacian-age body ornaments across Europe. In H. Knecht, A. Pike-Tay and R. White (eds), *Before Lascaux: the complex record of the early Upper Palaeolithic*, pp. 277–299. CRC Press, Boca Raton, FL.

Woodhouse, H. C. 1997. Ostrich eggshell beads in southern Africa. *Rock Art Research* 14: 41–43.

Ziegert, H. 2007. A new dawn for humanity: Lower Palaeolithic village life in Libya and Ethiopia. *Minerva* 18(4): 8–9.

Chapter 5
An Alternative Paradigm

A Sense of Perfection

The significance of human consciousness, expressed in an advanced level of self-awareness, to understanding how we became human hardly needs to be emphasized here; it seems self-evident. The centrality of art to the evolution of consciousness is not a new idea; it has been expressed by various writers (for instance, Read 1954: 143; Steiner 1964, 1970, 1972, 1990; Gebser 1985: 316–333, 487–505; Shiner 2001; Bednarik 2003) for many years. Whereas the replacement model, most strongly expressed in the "African Eve" hypothesis, postulates that there was no major change in human behavior until the "big bang of consciousness" of the Aurignacian (White 1995, 2003; Noble and Davidson 1996; Mithen 1999, 2004; Klein and Edgar 2002; Mellars 2005), the gradualist model derives greater support not only from the empirical evidence, but also from other disciplines, such as evolutionary anthropology and psychology (Lock and Peters 1996; Hodgson 2000, 2003; Bloom 2001; Dunbar 2003; Sedikides et al. 2006). For instance, the proposition of language emergence during the last third of the Late Pleistocene is widely rejected by nonarcheologists, with a minimum duration of 500 ka postulated (Dunbar 2003). This is consistent with the evidence of very early seafaring, which as we have seen demands quite complex "reflective" communication by around 1 Ma ago, also evident from Bickerton's (2010) hypothesis which at present offers the most plausible model of language origins. Sedikedes et al. (2006: 66) also perceive human self-awareness having been established before the appearance of archaic *Homo sapiens*, and in fact detect its "first glimmerings" "in the late stages of the *Homo ergaster/erectus* period".

One of the greatest fallacies of the short-range advocates is the claim that the available pre-Mode 4 evidence for symboling behavior is too rare, too sporadic to indicate any repetitive patterning. Precisely the opposite is true. The consistency of petroglyphs of Mode 3 traditions is remarkably uniform across all four continents occupied by such societies. In the Old World continents, the earliest rock art consists initially of cupules, together with a very few linear grooves. The next stage, from southern Africa (Fig. 5.1) to Australia (Fig. 5.2), features circle petroglyphs, either single or in certain specific types of arrangements, as well as a few other motif types. Images of tracks are added to this repertoire in Australia around the end of

R.G. Bednarik, *The Human Condition*, Developments in Primatology: Progress and Prospects, DOI 10.1007/978-1-4419-9353-3_5, © Springer Science+Business Media, LLC 2011

Fig. 5.1 Circle and cupule petroglyphs of the Middle Stone Age, Klipbak 1, Korannaberge, South Africa

the Pleistocene (Bednarik 1988), which is also the case in both North (e.g., Malotki 2010) and South America (e.g., Carden 2009), the continents apparently last colonized by humans. This uniformity is such a universal feature of Mode 3 technologies that the claim of the replacement people implies a stunning level of ignorance. After all, there are hundreds of thousands of Mode 3 petroglyphs in just two regions, Australia and sub-Saharan Africa, which has apparently never been noticed by those demanding "patterning." There is in fact far more uniform "patterning" in Mode 3 traditions than there is in Mode 4, with its wide divergence in repertoire.

Another example defeating the short-range model is the incredible longevity of some of these symbolic traditions, lasting literally for hundreds of millennia. Consider, for instance, the engraved forest elephant (*Palaeoloxodon antiquus*) bone from Bilzingsleben, Germany (Mania and Mania 1988; Bednarik 1995), named No. 1 (Fig. 3.12 top). It bears an arrangement of two distinctly grouped sets of subparallel engraved marks. White (1995) has claimed that these show no more patterning than he has found on his kitchen cutting board. Since then, Steguweit (1999) has demonstrated that the stone-tool cuts are not incidental; they were made deliberately. But what is particularly interesting is the discovery of another engraved bone at Oldisleben, also in Thuringia. Precisely the same grouping of two sets is present,

Fig. 5.2 Sacred Canyon, Flinders Ranges, South Australia: the lower and upper petroglyphs are of the Pleistocene, those on the diagonally exfoliated, inset middle panel are of the Holocene

with one set oblique to the other (Fig. 3.12 bottom). This time the intentionality of the maker is unmistakable, because of the way he or she deliberately placed each mark after first making a subsidiary cut to space the lines carefully (Bednarik 2006: 115). Despite the close geographical vicinity of the two find sites, they are separated by a couple of hundred thousand years. The megasite of Bilzingsleben is Lower Paleolithic and being of the Holstein interglacial complex in the order of 350–300 ka old (Schwarcz et al. 1988). The Oldisleben finds occur with Micoquian stone tools and although they lack secure dating, they are of the Eem and roughly in the order of 100 or 130 ka old. A comparison of the two marking strategies thus suggests an almost incredible longevity of the inherent patterning.

This "conservatism" is, however, not so surprising when we consider even more extreme examples. The durability of the tradition of creating cupules is unmatched in all of known rock art traditions. Cupules are hemispherical, cup-shaped depressions hammered into rock, in most cases with hand-held stonehammers. They are most frequently between 2 and 6 cm diameter, and they are usually as deep as it is possible to make them with such tools, keeping the diameter as small as possible at the same time. These petroglyphs occur most often in groups, and as demanded by taphonomic logic, the oldest specimens survived exclusively on rocks of hardness 7 on Moh's Scale: quartzites and even crystalline white quartzes, in some cases. The oldest cupules ever found, i.e., the oldest known rock art, date from an Oldowan-like stone tool industry in central India, comprising choppers and rough cobble tools. This typical Mode 1 industry was overlain by significant Acheulian (Mode 2) deposits at two cupule sites, Auditorium Cave (Bednarik 1993) and Daraki-Chattan (Bednarik et al. 2005). The second site yielded not only exfoliated rock fragments with cupules from all sediment layers, but also contained in the handaxe-free Mode 1 horizon numerous of the hammerstones that had been used in creating the site's 530 cupules. The age of this deposit still needs to be established securely, but it is expected to be several hundred millennia.

Cupules are also the only rock art apparently surviving from the Fauresmith tradition of southern Africa, which some see as a late variant of the Acheulian, others as an early Middle Stone Age tradition of small, relatively refined handaxes. It is often 200–300 ka old, although much earlier estimates are also available. Again, the period's cupules occur exclusively on fully metamorphosed quartzite facies, having been identified at two sites in the Korannaberg region and nearby (Beaumont and Bednarik 2010), in the southern Kalahari (Fig. 5.3). Another African occurrence of the phenomenon is the slab from Sai Island, Sudan, also of the late Middle Pleistocene (about 200 ka old). The early cupule tradition can even be found in Europe, with the classical specimen from La Ferrassie, the sepulchral block over "Neanderthal" burial No. 6 being the oldest example (Fig. 5.4). This is considerably younger, of the second half of the Late Pleistocene. Finally, there are hundreds, if not thousands, of examples of Mode 3 cupule panels in Australia. Cupules in fact continue into Mode 4 traditions, especially in Eurasia, and they are found in numerous traditions of the Holocene, from Africa to Australia, from the "Mesolithic" to the metal ages, even from medieval times in Europe; while in Australia they were still produced in the twentieth century (for comprehensive review, see Bednarik 2008a).

Fig. 5.3 Cupules at water hole, Nchwaneng, southern Kalahari, Fauresmith or Middle Stone Age

Fig. 5.4 The sepulchral block with cupules placed over 'Neanderthal' child's burial, La Ferrassie, France

The production of cupules is an entirely "pointless" pursuit, it has no utilitarian purpose whatsoever—as we know from the few ethnographic accounts we have. These petroglyphs were made for several hundred thousand years—longevity of patterned symbolic activity unmatched in the history of this planet.

A productive direction of investigation would be to examine the behavioral aspects of cupule production. It has been studied extensively through a series of replication experiments, which have yielded surprisingly consistent empirical information about these strange features. Not only do they occur very widely in all continents except Antarctica, the technological intention evidenced in their statistics also seems to be of incredible uniformity—especially in relation to the early traditions. Replication demonstrates that there was a very deliberate convention to keep the diameters of cupules as small as technically possible. Those made on particularly hard rocks often required upward of 30,000 blows with hammerstones, as noted in Chapter 3, many of which would have been worn out in the process of creating just one cupule. When made on the softest available rocks (e.g., Miocene limestones), cupules tend to be far deeper than wide, and there is a clear endeavor to keep the opening as small as possible (Bednarik 2008a). Not only is this pattern practically worldwide, emphasizing the nonutilitarian nature of the phenomenon, it hints at a very specific behavior pattern. Since cupules occur often in large groups densely covering a specific rock panel (Fig. 5.5), there is a distinct impression that the convention was based on a culturally determined ideal of a perfect cupule: maximum achievable depth at minimal opening size. We have not the faintest idea why these characteristics were pursued with such extreme dedication, with such a relentless thirst for perfection. Would we be justified in suspecting a connection between this pointless—in a Darwinian sense—pursuit of an ideal and the dominating factor in our own lives: objectives and quests that offer no apparent Darwinian advantages?

The sense of perfection apparently evident in cupules, occurring as far back as the Mode 1 industries of the "Lower Paleolithic," is of great relevance here, because such a faculty seems to have no justification unless there is a distinctive state of self-awareness. The enormous investment of energy in such "senseless" productions as cupules needs to be explained, particularly as it coincides with other such evidence from that early period. For instance, detailed replicative experiments in making ostrich eggshell beads matching precisely those found in the "Late Acheulian" of the shores of former Lake Fezzan, Libya, have also led to fascinating behavioral deductions (Fig. 3.15 bottom) (Bednarik 1997). It was found that their size as well as the perfect centrality of their perforation indicates a very deliberate practice. The smallest size such a bead can realistically be ground down to is about 6 mm diameter (the *heishi* technique was unknown in the Acheulian; cf. Francis 1989). As the size approaches this order of magnitude, the disc becomes increasingly difficult

Fig. 5.5 Cave wall with cupules, Mode 1 Lower Paleolithic tradition, Daraki-Chattan, India

to hold between fingers, and as the fingertips begin to rub against the grindstone with reducing bead size, the process becomes rather painful. Also, since the diameter of the stone drill-made central hole can be no smaller than 1.4–2.0 mm, it follows that the bead's fragility increases exponentially as the outside diameter of 6 mm is approached (I tested this proposition empirically). Therefore, this diameter represents the smallest size at which the bead remains structurally strong enough to withstand *some* rough handling.

Moreover, the Acheulian beads are very well made, with a near-perfect circular outer margin and an equally perfect rim thickness all around (Fig. 3.15). These precise forms can be achieved only intentionally, by constant checking of the shape during the final abrading phase. It is nearly impossible, according to replication, to obtain such a perfect round shape and centrality of the perforation by accident. This means that the makers had not just a well-developed sense of symmetry, but also a clearly defined concept of the perfect geometric form they aspired to—a form Lower Paleolithic hominins would have had few conceptual models for to underpin as a mental template.

Even if it is preferred to have a perforated bead (which is not necessary, as many later, merely notched pendants show), this does not necessarily call for a *central* perforation. The rational explanation why the maker would go to such lengths to abrade the bead equidistantly is the involvement of a sense of perfection. This proposition is confirmed by the size of the beads; it seems self-defeating to make beads as small as technically possible, 6 mm diameter. If it were the purpose of a bead to be seen, a large bead would fulfill that role much better. Yet the (measured) labor investment of making a very small bead is significantly greater than that required for a large bead. There is a palpable impression that the primary objective was to push the available technology to its very limits, creating a geometrically perfect form. It is from this perspective that we need to examine these concept-mediated (Bednarik 1995), symbolic objects. They were imbued with meaning, they were exograms (see Chapter 6), and their laboriously achieved perfection expressed social values we can reasonably assume entailed benefits for the maker or wearer.

A sense of perfection may also be evident in the unnecessarily refined stone tools, especially "handaxes," of the Late Acheulian, which have often been suggested to involve nonutilitarian dimensions of the same kind. Any of these examples of incredible efforts and apparent motivation, especially in the case of many of the cupules, demand the existence of abstract values that were effectively understood by the beholder. That applies of course to any society that uses objects such as beads or pendants: the very existence of such objects virtually presupposes the use of complex symbolisms, indicating essentially cognitive and cultural modernity in Middle Pleistocene hominins.

The Domestication of Humans

These considerations are of much relevance to an informed behavioral science of hominins and yet, because of latent censorship by the dominant paradigm of Pleistocene archeology, they have not before been available to the scientific (as opposed to archeological) study of evolutionary processes. The significant shortcomings of Pleistocene archeology we have already visited are in some ways mirrored in those of socio-cultural anthropology, another anthropocentric discipline gradually being displaced by the behavioral sciences and evolutionary biology. Just as archeology is adrift without a universal theory, subject to the vagaries of accidental discoveries and chasing its own hermeneutic tail, social and cultural anthropology "has been much less demonstrably productive [than evolutionary biology] over the same period" (Mesoudi et al. 2006: 329–330). This is hardly surprising; after all, evolutionary biology has a very clear underlying theory, that of neo-Darwinism. Ultimately, all "social sciences" will have to yield to the sciences, but the precise composition of a science of the evolution of human behavior remains the subject of much debate, and it should not be determined solely by neo-Darwinism. As so often in the academic project, reality is a great deal more complex than heuristic rationalization envisages—as we shall soon see.

Evolutionary theory attributes evolutionary change essentially to two factors, natural selection and sexual selection. In the first, specific phenotypes representing aspects of morphology or behavior are preferentially reproduced across generations of a given population. In the second, phenotypes become over-represented, either through mate choice or intrasexual competition. The emphasis is on genetic inheritance, although the hard evidence for this is no more secure than the genetic basis of the doomed African Eve hypothesis.

Over recent years several new directions of inquiry have evolved, challenging simplistic evolutionary theory. Developmental systems theory replaces the overly restrictive focus on the genes with a model of interacting systems (Oyama 2000; Oyama et al. 2001). While vague overall, it does raise some pertinent points, especially concerning the nongenetic inheritance of traits and the cybernetic feedback from organism-environment systems changing over time. Niche construction has been presented as another major force of evolution (Odling-Smee et al. 2003), operating similar to natural selection. In rather the same way as we have in the first chapter seen visual and mental taxonomizing processes and the inclusion of new neural structures becoming available for evolutionary selection in feedback systems, niche construction also creates feedback within the evolutionary dynamic. Organisms engaged in it modify the evolutionary pressures acting on them, as well as on other but unrelated populations sharing the same space. Humans are rightly seen as the "ultimate niche constructors" in which their complex cultures play an important role. Laland et al. (2000) see much of niche construction as guided by socially learned knowledge and cultural inheritance (cf. Silk 2007).

The perhaps most effective niche created by hominins is their language (Bickerton 2010: 11, 12, 219–221). The most plausible explanation of how human language began presents it as deriving from "power scavenging," i.e., the exploitation of carcasses by driving off competitor carnivores, through the need for "recruitment" of conspecifics (Bickerton 2010). Based on the observation that about two million years ago, taphonomic evidence begins to show that tool marks on bone precede rather than postdate carnivore teeth marks, Bickerton (p. 125) posits the need for communicating the discovery of carcasses to recruit helpers to drive off other species, which would involve "displacement," a facility absent in all animal communication system (pp. 160, 216, 217). Displacement defines the ability of referring to entities not immediately present in time or space. Bickerton (2010: 231) rightly emphasizes that niche construction and language production are both autocatalytic processes: once started, they drive themselves, creating and fulfilling their own demands.

Evolution has been suggested to encompass also other "dimensions," termed epigenetic, behavioral, and symbolic inheritance systems (Jablonka and Lamb 2005). All organisms are said to be subject to epigenetic inheritance, which refers to physiological/biological process above the level of DNA. Behavioral inheritance is found in most species, and defines the transference of information or behavior through learning rather than genetically. Symbolic inheritance, with which we are primarily concerned here, is apparently found only in humans. The underlying contention of these new ways of thinking is that evolution is not a simple genetic process relying on the appearance of mutations. The idea that human evolution simply cannot be assumed to have been a purely biological process is not at all new (Dobzhansky 1962: 18, 1972). It has recently received a new impetus from increasingly sophisticated work, and the notion of a progressive moderation of human evolution by culture is the central plank of the gene-culture co-evolutionary model (Boyd and Richerson 2005; Richerson and Boyd 2005). Most recently, Fuentes (2009) has sought to reconcile the pronounced duality of evolutionary biology and socio-cultural anthropology, pointing out that symbolic and other cultural processes influence behavior and potentially physiological and even genetic factors. His demand that behavioral plasticity has a specific role in human behavior runs again counter to neo-Darwinism, but it seems impossible to explain hominin development, especially of the Late Pleistocene, without that factor.

These new developments are certainly useful, especially in that they reject the role of genetics in "explaining everything" in hominin evolution. They also express considerable criticism for the self-confirming paradigms of recent decades, critique that is so crucial to a sound epistemology. One does not need to subscribe to pan-critical rationalism, a philosophy that rejects the existence of any authority, proof, disproof, justification, probability, and is even critical of observation, logic, and its own position and criticism. The debilitating, all-pervading appeal to authority governing archeology does need to be severely challenged, and this has not occurred adequately. However, there are two significant shortcomings of these various strands of criticisms coming from the sciences. One is that they have not produced an alternative paradigm; they have merely illustrated problems that need to be attended to. The other concerns the lack of relevant empirical evidence, which the sciences simply have no access to because archeology is either itself unaware of its existence (as far too often appears to be the case); or alternatively it has made great efforts to discredit such evidence in order to uphold its dogma. Therefore, the position of the behavioral, cognitive, and semiotic sciences is essentially that they have detected flaws in the dominant model of the emergence of human modernity, but they are not in a position to offer an alternative: archeology dominates the discourse on hominin evolution.

The scenario remains that there is a significant change in the physiology of humans during the last 50,000 years in Europe, and modern Europeans differ genetically from robust Europeans 50 ka ago. The same change from Robusts to Graciles occurs in other continents. Not only do these changes need to be explained, there is another issue which, oddly enough, the replacement advocates are silent on: the changes that did occur contradict all canons of Darwinian evolution. Without a significant change in their environmental niche, these humans experienced numerous deleterious physiological changes to become gracile (as well as deleterious neurological changes, as we will see in Chapter 7). The thickness of their skulls decreased radically, as did the general robusticity of their skeletons. The traces of muscle attachments indicate that physical strength declined markedly, perhaps by as much as 50%. On top of that, their brain shrank by around 200 ml, and that took place during a time when we think the demands on their mental abilities increased exponentially. These changes are certainly dramatic, occurring in fact over just a few tens of millennia. In the history of the human genus, there is no evidence of such rapid changes, and conventional wisdom has it that all previous changes were for the better of the species concerned. That certainly cannot be said about what happened in the most recent history of human evolution, which in fact looks much more like devolution, or evolution in reverse.

So what happened? If it was not a case of invasion by physically (and perhaps even intellectually) *inferior* Africans of evenly matched technology, what alternative is there? The answer is provided by a combination of two strands of determinants. One is the indisputably very major influence sexual selection has on who passes on their inheritance; the other is the rising power of cultural imperatives over natural. When breeding mate selection becomes moderated by cultural factors (such as cultural constructs of attractiveness, along with perhaps social position, communication ability, body adornment), the laws of evolutionary theory become suspended, and are supplanted by Mendelian laws of inheritance (Mendel 1866), the basis of the discipline of genetics: evolution by natural selection is replaced by breeding, or artificial selection, resulting in *domestication*. It was in studying artificial selection in pigeons that Darwin detected the similarity with natural selection, and here at last we seem to come full circle: modern humans are the result of incidental self-domestication. Here we will discover that, in their fetish of using the purported travel of genes to infer the movement of major populations, Pleistocene archeologists ignored that it is not evolutionary genetics that determines inheritance: in the end even Darwin has to defer to Mendel.

This revolutionary alternative has been outlined as an option to the replacement hypothesis (Bednarik 2007, 2008b, c, d) and here it is explained in some more detail. The apparently most important question to be asked in this context is this: what could have caused the inherent laws of biological evolution to be suspended for humans during the last fifty millennia or so?

It is particularly important to note that the change seems to have occurred universally and roughly concurrently, in all four continents occupied by hominins by 50 ka BP. Since this enormous geographical range involved numerous climatically and environmentally different niches, from the tropics to the Arctic, it is impossible to explain such largely uniform change from robust to gracile as the result of natural selection. The same rejection of evolutionary dynamics may be implied by the relatively swift conversion, taking only a few tens of millennia. In southern and eastern Europe, one might argue that the Campagnian Ignimbrite Event and subsequent sharp climatic decline almost exactly 40 ka ago (Fedele and Giaccio 2007) may have precipitated demographic and cultural adjustments. Although this environmental bottleneck could have effected genetic or anatomical changes in some parts of Europe, there is no evidence that it did, and the universal human gracilization over the last 50 ka or so demands a universal explanation and precludes a local one. Occurring concurrently in the course of the second half of the Late Pleistocene, in all four continents occupied, this process needs to be explained if we are to understand our origins.

In Europe, it is best documented by human remains from the central region, particularly in the Czech Republic, from the crucial period of about 31 ka to 26 ka BP, which witnessed distinctive sexual dimorphism. Despite the lack of credible stratigraphic evidence from the Mladeč site, the recent attempt to provide direct dates from some of its human remains suggests that they represent precisely this interval (Wild et al. 2005). As we have seen in Chapter 2, male crania are characterized by typical robust features. As in "Neanderthals," cranial capacities exceed those of "anatomically modern humans," but we noted that there is a reduction in the difference between male and female brain size relative to Neanderthal data (Fig. 5.6). The female specimens show similarities with, as well as differences from, accepted Neanderthal females, but are far more gracile than the males, while still being more robust than males of later Pleistocene periods. The Mladeč population as well as contemporary others in central Europe (e.g., Pavlov Hill, Předmostí, Dolní Vestonice, Podbaba, Miesslingtal) thus seem to occupy an intermediate position between late Neanderthaloid *Homo sapiens* and *H. sapiens sapiens*.

Gracilization begins typically in females, with males lagging many millennia behind (Fig. 5.7). The process has continued to the Holocene, and reduction in both dimorphism and robusticity is also still active in human "evolution" today. The face, jaw, and teeth of European humans 10 ka ago are in general 10% more robust than

Fig. 5.6 Mladeč Cave 1, 6, and 5, Czech Republic, showing the striking morphological differences between the two females on the *left* and the male on the *right*. (To facilitate comparison, all specimens are shown facing the same direction.)

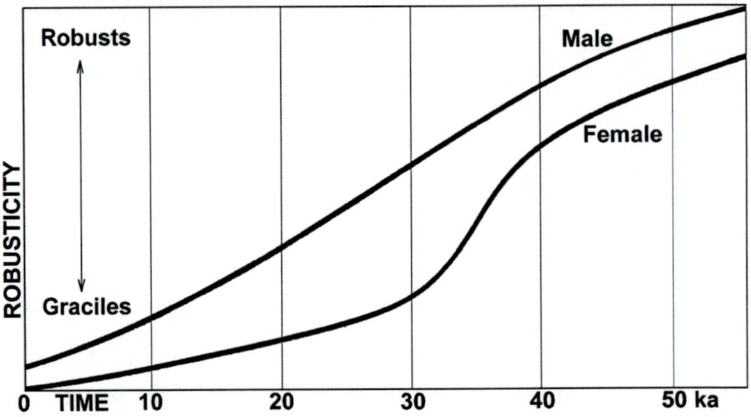

Fig. 5.7 Male and female relative cranial gracility in Europe through time, showing that the decline in robusticity is gradual in males, but accelerated in females between 40 and 30 ka BP

those of today's Europeans (and Asians), and those of 30 ka ago are 20–30% more robust. Some modern humans (e.g., Aborigines) have retained tooth sizes typical of archaic *H. sapiens* and other robust features are preserved in many populations or individuals. Neanderthaloid specimens occur in the "Mesolithic," such as the Hahnöfersand specimen already mentioned, or the equally robust "Mesolithic" skull fragment from Drigge, also from northern Germany, which is about 6250 years old (Terberger 1998). Numerous other late specimens of Robusts range in age from the Magdalenian through to the Neolithic, and younger.

Holocene gracilization could conceivably be explained as a response to changing food-processing techniques or less physically demanding lives. The smallest tooth sizes tend to be found in those areas where food-processing techniques have been used for the longest time. However, this explanation cannot be extended to universal gracilization during the Late Pleistocene. The life style of people 15 ka ago is not thought to have been significantly different from that of 35 ka ago, yet the overall rate of gracilization appears to be have been reasonably uniform over the past 50 ka (Fig. 5.7). As a universal phenomenon, it has not been explained, and indeed has been ignored due to the dominance of the replacement model.

Natural selection simply cannot account for a significant reduction in robusticity and reversal of encephalization without any apparent trade-off in evolutionary benefits for the organism in question. No such benefits are apparent, and yet this process seems to have been universal wherever humans existed during the Final Pleistocene. It is proposed here that the dimorphism observed during the crucial period of the last twenty or thirty millennia of the Pleistocene presents the key to the most parsimonious explanation. Dimorphism in mammals generally reflects one or both of two selection pressures: competition between males for access to females, or male–female differences in food procuring strategies, with males provisioning females (Aiello and Wheeler 1995; Biesele 1993; Deacon 1997). In the case of late hominins it has been suggested that physical competition among males may have been diminished radically with the introduction of accurate projectile weapons acting as "equalizers" (Boehm 1993, 1999). This is, however, not a satisfactory explanation: effective distance weapons were in use long before the Upper Paleolithic (as noted in the previous chapter, spears of the Lower Paleolithic were found at seven European sites), together with evidence of large game hunting. Thus, the "equalizers" had long been in use and they do not explain the gender-specific pattern of later gracilization, nor the extensive fetalization that took place in the Final Pleistocene.

The Fetalization of Humans

Human evolution, particularly in the latter part of the Pleistocene, simply cannot be assumed to have been a purely biological process (Dobzhansky 1962: 18); it must have been increasingly moderated by culture, as predicted by gene-culture co-evolutionary models. It is self-evident that practices such as deliberate breeding-mate choice determine procreational success today (Buss and Barnes 1986; Buss et al. 1990), so the obvious question to be asked is: at what point in time did it first appear? Other primates (indeed, all other animals) exhibit no preferences in mate selection of youth or specific body ratios, facial features, skin tone or hair; yet in present humans these are deeply entrenched, perhaps "hardwired." Facial symmetry, seen to imply high immunocompetence (Grammer and Thornhill 1994; Shackelford and Larsen 1997), is also of importance, and in female humans neotenous facial features are strongly preferred by males (Jones 1995, 1996). Since this applies undeniably today, the rational way to examine this issue is to consider at what point in human development the influence of nonevolutionary currents can be detected. It is suggested here that around 40 ka ago, cultural practice had become such a determining force in human society that breeding mate selection became increasingly moderated by cultural factors, i.e., by factors attributable to learned behavior. These could have included the application of a variety of cultural constructs in choices such as social standing, communication skills, body decoration (which becomes notably prominent 40 ka ago), and most especially *culturally negotiated constructs of physical attractiveness*.

In all animals, including all hominins, reproductive success determines phylogenetic direction. It is obvious that today, the processes of natural evolution are largely suspended in our species' development, having been widely replaced by cultural mating imperatives. Inescapably, this development must have been phased in at some time in our past. If we were to look for evidence of its timing, two strategies spring to mind. We could look for signs that attributes of natural fitness were decisively replaced by attributes that confer no Darwinian survival benefits, or we could look for indications of a culturally mediated preoccupation with female sexuality. We would note that, firstly, gracility, especially of females, develops strongly during the "Aurignacian" and "Gravettian"; and secondly, that this very same period is marked by a distinctive preoccupation with female sexual attributes. The latter is found in the common depictions of (mostly) isolated vulvae or pubic triangles; for example, at Abris Blanchard, Castanet, Cellier and Le Poisson, La Ferrassie, Laussel (Delluc and Delluc 1978) and in Chauvet Cave; and the creation of naturalistic female statuettes, often with pronounced sexual aspects, beginning with the "Aurignacian" (Bednarik 1989; Conard 2009). Therefore, the question to be asked is: what cultural preferences could possibly have led to the gracilization of female humans during the second half of the Würm glacial in Europe?

Mating preferences and their genetic results in respect of personality and anatomical traits (Laland 1994), which could become cultural selection variables, can be modeled by methods of the gene-culture co-evolutionary model (Cavalli-Sforza and Feldman 1973; Feldman and Cavalli-Sforza 1989; Aoki and Feldman 1991; Durham 1991). It has been noted that traits selected for can include large female breasts, small feet or male macho behavior, and most certainly physical "attractiveness"—informed no doubt by cultural constructs of attractiveness (such constructs are not objective, they exist purely in the human mind; Barkow 2001). The question then becomes: if the recent gracilization of humans were related to fetalization, what would be its anatomical consequences?

Humans resemble chimpanzees anatomically most closely in the latter's fetal stage (Haldane 1932; De Beer 1940; Ashley-Montagu 1960). Both the fetal chimpanzee and the adult human have hair on the top of the head and on the chin, but are otherwise largely naked. In apes, this changes rapidly upon birth, in humans it remains for life. All male adult apes have a penis bone, but it is categorically absent in both fetal chimpanzees and all male humans, from the fetal stage and throughout life. In fact the penis bone of apes is one of the very last parts of the ape fetus to form, shortly before birth, and its atrophy in humans appears to have been compensated for by significantly increased penis length and thickness, relative to apes (Badcock 1980: 47). Similarly, in female chimpanzees, the *labia majora* are an infantile feature; in humans they are retained for life. The hymen, too, is present only in the neonate ape, but is retained for life in human females in the absence of penetration. The organs of the lower abdomen, such as rectum, urethra, and vagina, are typically aligned with the spine in most adult mammals, including apes; only in fetal apes and humans do they point forward relative to the spine (upright walking appears irrelevant, because fetal apes do not walk). The human ovary reaches full size at the age of five, which is the age of sexual maturity of the apes (De Beer 1940: 75). Most importantly, the skull of an unborn ape is thin-walled, globular and lacks the prominent tori of the adult ape, thus resembling the cranium of a modern human. Upon birth its robust features develop rapidly. The slow closing of the cranial sutures in humans is again clearly a neotenous feature (genes RUNX2 and CBRA1 refer). The face of the ape embryo forms an almost vertical plane, as it does in the modern human all the way through adulthood, which is certainly not the case in mature apes. Even the brains of fetal apes and adult humans are much more similar to each other, in terms of proportion and morphology, than they are to those of adult apes.

These and many other features define the anatomical relationship between ape and man as the latter's *neoteny*. The legs of fetal apes are relatively short, while the arms are about as long in relation to the body as in humans. In the apes, the arms become much longer after birth. Human hands and feet resemble those of embryonic apes closely, but differ significantly from both hands and feet of mature apes. In fact the human foot, especially, retains the general structure found in unborn apes, which rather contradicts the hypothesis that it is an adaptation to upright walking. It could equally well be the case that upright walk is an adaptation to the neotenous foot of hominins. Even the shape of the cartilage of the ear in humans is a neotenous feature.

In neoteny, sexual maturity is attained before full somatic development, and juvenile characteristics are retained for life. In an evolutionary perspective, it refers to species whose adults retain juvenile ancestral features. This has also been called fetalization, because in such phylogenic development, fetal characteristics remain into adulthood, and specific processes of anatomical maturation are retarded (De Beer 1940). Indeed, it is fascinating to note that in human fetalization, biological history seems to be repeating itself: all vertebrates appear to be the result of neoteny in chordates (species having a notochord) hundreds of millions of years ago (De Beer 1940: 76–78). The modern human has undergone so much selection in favor of neoteny that this retardation should be seen as rivaling in importance the distinguishing anatomical characteristic of the oversized brain. It, therefore, needs to be considered here. "But neoteny does not only contribute to the production of large structural change; it is also the cause of the retention of plasticity" or "morphological evolvability" (De Beer 1930: 93). Adaptively useful novelties supposedly become available as maturation genes are freed by pedomorphosis. This neotenous "retention of plasticity," also noted by Fuentes (2009), could be a key factor in how humans became what they are; here it will need to be considered in more detail.

Encephalization and neoteny in hominin evolution are quite probably related, perhaps through supervenience. It is self-evident that, relative to the neonate ape, the newborn human is not remotely as far developed. For instance, it would find it impossible, for many months after birth, to cling to the fur of a mother for transport. Of course this is related to its excessive brain size, which has caused it to be expelled at a much earlier stage of fetal development. It can be regarded as highly probable that human mothers always had to carry their infants. Indeed, one of the first kinds of artifacts used by early humans was probably some kind of sling or baby-carrying bag. The long period during which the human infant was entirely dependent upon the mother, not just for sustenance but also to move with the horde as well as for protection, extended the period for learning very significantly. This, obviously, coincided with the continued growth of the brain after birth, which in fact exceeds that of the fetus in man. In the first year after birth, our brain more than doubles in both volume and weight. It continues to grow, approaching adult size by the age of three, but goes on expanding slightly more up to adolescence and even beyond. If we compare this extraordinary development, unheard of in the rest of the animal kingdom, with that of other primates, we see that in simians such as the rhesus monkey and in the gibbon, 70% of adult brain size is achieved at the time of birth, the remaining 30% in the subsequent six months. In the larger apes, the size of the brain approaches adult size after the first year of life. These are very significant differences, and they are all connected with our neoteny.

Another marked difference between humans and most other animals is the abolition of estrus, or periodicity of libido in the female. This almost uniquely human feature has not been explained satisfactorily, but there is a good probability that it is also related to these factors—through one of two alternative scenarios. The excessively long period of infant dependency would have been mirrored in a similar dependency of mothers on the horde, most especially for the meat protein needed for the brain tissue of their unborn (Aiello and Wheeler 1995; Leonard 2002; Leonard and Robertson 1992, 1994, 1997). It is thought probable that there was strong selection favoring female mutations allowing long periods of sexual receptivity, leading to the abandonment of estrus altogether: those females who were longer or always receptive were favored in the distribution of meat from kills, in a feedback system facilitating fetal encephalization through better access to animal protein (Biesele 1993; Deacon 1997). It has been noted that on occasion, female chimpanzees are only given meat after they have copulated with a successful hunter, and it is logical that such a behavior trait would select in favor of continuously receptive females.

The second alternative explanation for the loss of estrus in humans is simpler and would favor a very late introduction, but may seem no more than a stab into the dark. Domesticated mammalian species lack the seasonal reproduction of their wild ancestors and most can reproduce themselves at almost any time of the year. It is remotely possible that the same effect in humans is the result of their self-domestication.

Be that as it may, the numerous physiological features of human neoteny should suffice to demonstrate that humans are anatomically best defined as a fetalized form of ape. Although the process of selecting in favor of infantile physiology appears to mark much of human history, during the Final Pleistocene it suddenly accelerated to an unprecedented rate and resulted in markedly unfavorable mutations, from the perspective of natural selection. The brain of all higher species is hard-wired to react in a nurturing fashion to neonate features, but this does not seem to result in fetalization unless moderated by culture. Worldwide, wherever humans existed 40 or 50 ka ago, possessing as they did Mode 3 technological traditions, they shed all of their robust features in just a few tens of millennia. Their brain size decreased, despite the rapidly growing demands made on their brains. Their muscle bulk waned until their physical strength was perhaps halved, in tandem with significant reductions in bone strength and thickness. The decrease in skull thickness is particularly prominent, as well as relatively rapid reduction in cranial robusticity. This process occurred so fast that it can be tracked through the millennia. At about 35 ka, we encounter partially gracile specimens from Europe to Australia. The subsequent skeletal evidence presents a distinctive sexual dimorphism: the female crania, though still much more robust than male crania were toward the end of the Pleistocene, show distinctive gracilization (development of globular crania, reduction or absence of supraorbital tori and occipital projection, significant loss in bone thickness, reduction in prognathism, and several other features bringing humans closer to the fetal ape's morphology). The males, however, remain almost as robust as typical "Neanderthals." Ten thousand years later, the females have become markedly more gracile, and the robust features of the males have also begun to wane. Toward the end of the Pleistocene, the males begin to catch up with the females, and from there on the loss of robusticity continues right to the present time (Fig. 5.7).

The Pain of a New Idea

The explanation proposed here is radically different from any other so far offered for the key phenomena discussed in this book. It is an idea I believe has not been expressed by anyone previously. The closest notion I have only recently found is Eugen Fischer's (1914) objectionable suggestion of almost a century ago that "races" may have been subjected to "breeding." Firstly, the whole idea of racialism (Brace 1996, 2005) is misguided and baseless; second, Fischer happened to be a persistently racist commentator, closely associated with a fascist regime.

In all sexually reproducing species, all characteristics of individuals are said to be inherited through genes. It does not necessarily follow that all inheritance must be encoded in DNA. The principles and mechanisms of genetics apply to the molecular structure of cells and tissues, the development of individuals and the evolution of whole populations. Selective breeding defies natural evolution in the sense that it can rapidly change the characteristics of a population without any natural selection in the Darwinian sense.

To appreciate the effectiveness of such a process we only need to recall the dramatic example of the effects of domestication provided by the dog, bred in just 15 ka to far more radical skeletal extremes than are evident among all hominins for more than 3 million years. Given the right selective factors, domestication can occur rapidly rather than gradually. Importantly, domestication demonstrates that the continuous selection of a single trait does not necessarily evolve a population of better-adapted organisms, as natural selection would be expected to yield. Rather, it shows that selection for a single trait results in changes in numerous traits, changes that are usually deleterious, be they physiological or a range of others. For instance, domestication of animals typically results in decreased cranial volume relative to body size, a decrease that in some species (e.g., pigs) can be as much as 30–40%. In hominins, we have seen that the reduction in brain size over just a few tens of millennia amounts to at least 200 ml. Leading physical anthropologist Henneberg (1990) comments on the reduction of human cranial capacity since the MSA. When we see an *increase* by that amount in the earliest hominins we regard it as significant and wax lyrical about encephalization, even though it required many hundreds of millennia of evolution. But when such a staggering loss of volume occurs in a relatively short time, it is ignored so as not to question the replacement dogma.

We would be hard pressed to deny that cultural determinants are powerful in the choices we make *today*; therefore, there must be a point in time when these began to override Darwinian selection. One could argue that this "evolutionary luxury" perhaps occurred when humans developed the ability of producing staple food surpluses, through the advent of agriculture. Alternatively, we could look at the hard evidence and search for signs of phylogenic developments that defy natural evolution. I have chosen the latter here, and propose that the apparent reduction in evolutionary fitness evident in the recent neotenous gracilization of *Homo sapiens* marks the time when physical appearance became a cultural construct affecting mate choice. It began with a sexual male preference of females with mutations presenting juvenile somatic characteristics, i.e., females pioneered it. The decline of robusticity in males lagged many millennia behind the gracilization of females. Individuals considered attractive simply had more (and perhaps more successful?) offspring, *and it is they who "replaced" the robust genes.*

The paleoanthropological record worldwide suggests the gradual introduction of selective breeding between 50 and 30 ka BP. It is reasonable to assume that much of what constitutes sexual attractiveness is attributable to cultural constructs, although there may well be biological bases for these (such as immunocompetence, or the greater reproductive potential of young females because they offer more fertile years). Once these affect "conscious" mating choices, breeding patterns favor their perpetuation, and the population should be considered to experience domestication.

We have three basic hypotheses to account for the universal change from Robusts to Graciles in the final Pleistocene: replacement by an invading population in four continents (for which we lack any evidence, be it skeletal, technological, cultural or genetic); gene flow and introgression without any mass movement of population (which is demographically more plausible but fails to explain the apparent suspension of evolutionary canons); or cultural moderation of breeding patterns (i.e., domestication). Only the last-named option can account for all the hard evidence as it currently stands. With breeding mate selection becoming increasingly moderated by cultural factors we have a far more effective explanation for the worldwide precipitous change from robust to gracile types than any other offered so far. This is certainly not a development unique to Europe, it is found in Australia, Asia, and Africa as well. It is incumbent upon us to explain why a species should suddenly, in evolutionary terms, develop such regressive features as thinner skulls, smaller brains, significantly reduced bone and muscle strength, and perhaps even hair loss in a cold region. Nature does not select for such plainly disadvantageous variables, but culture might. The most logical explanation is that cultural factors had begun to dominate breeding patterns to the extent that *anatomically modern humans are the outcome of their own domestication.*

This demands a revolutionary change in the way we view hominin development in the last part of the Pleistocene. The skeletal evidence from central Europe suggests that this process began with males developing a reproductive preference for females of juvenile characteristics, whose genetic success only needed to be very marginally greater to achieve the changes the skeletal record documents. A selection bias of much less than 1% per generation suffices to yield the change observed. One of the most decisive issues is that in no animal species, right up to apes, do any of the factors determining modern human sexual preferences play any role: any bonobo will happily couple with any other, irrespective of facial symmetry, hair color, youth or whatever. It is, therefore, inescapable that ideas of a sexual desirability that was unrelated to mere reproduction must have been introduced at a certain time in human history. As they became reified (and apparently expressed in paleoart since 40 ka ago; Conard 2009; Porr 2010), their effect on breeding patterns would easily account for the progressive female gracilization we observe. This is then a case of cultural selection for specific phenotypes of juvenile features. Eventually, it also affected the male genotypes, resulting in the reduction of male robusticity that becomes marked during the "Gravettian" and continues to the present time. In short, humans "domesticated" themselves, unintentionally, well before they did the same with other species.

In general the domestication of animals and plants consists of the collective genetic alteration of their physiology, behavior or life cycle through selective breeding. Historically, this process is thought to have begun with the gradual and probably unintentional domestication of the wolf in the Final Pleistocene (possibly in the order of 15 ka ago), and the domestication of numerous plants and animals in the course of the Holocene. In general, the term has been used for such alterations caused, deliberately or not, by humans, but it has not so far been applied to humans themselves. In fact humans are often considered to be the initiators of domestication, even though there are numerous examples of domestication by other species. Many animal species, vertebrate and invertebrate, from mammals to ants, have domesticated others, for instance, to modify foods undigestible by the domesticators, or for their labor or simply to serve as staple food source. Humans, of course, are animals too, but to what extent they might be the product of their own "domestication" has not been the subject of any attention so far.

We can expect this new hypothesis accounting for the origins of "modern humans" to be contested most vigorously, especially by those whose hypotheses it obviates. Competition between scientific paradigms is not the sort of battle that can be resolved by proofs (Kuhn 1962). As Max Planck has pointed out, "[a] new scientific truth does not triumph by convincing its opponents and making them see the light, but rather because its opponents eventually die, and a new generation grows up that is familiar with it". Thus, paradigms are gradually replaced as more practitioners, especially the younger and those of more flexible thinking, are won over by the better explanation, until only a few elderly of strong allegiances remain. (Thus, even in the competition of hypotheses, De Beer's youthful plasticity and morphological evolvability win out in the long term.) It is unfortunately true that the acceptance of a model is not a function of its objective merits, but is a largely democratic issue. Typically, a new paradigm appeals to the individual's sense of the appropriate or the aesthetic; it is said to be neater, more suitable, simpler, or more elegant (Kuhn 1962: 155). While all of these apply to the domestication hypothesis, when comparing it to the awkward replacement hypothesis with its countless contradictions, it must be remembered that the proponents of an established paradigm have devoted their lives and careers to their cause and are understandably reluctant to relinquish their preeminence. That will certainly apply to the African Eve advocates, Pleistocene archeology being the sluggish discipline that it is.

The acceptance of a new paradigm entails also the renunciation of many of the books and articles in which that paradigm has been embodied. Academic textbooks are the pedagogic vehicles for the perpetuation of what Thomas Kuhn defines as "normal science." According to him, they are systematically substituted for the creative scientific literature that made them possible; they are immensely effective in the educational initiation of scholars. Yet they are inevitable abstracted, dumbed-down and diluted rationalizations of the often very complex concepts they popularize for academic mass-consumption. In this light the quasi-democratic operation of disciplines facilitates the perpetuation of dominant paradigms, and the domestication hypothesis, certainly a case of Kuhn's revolutionary science, is assured a rough ride for some time to come. That will not be a reflection of its weakness, and if it is a better approximation of reality it can be expected to prevail ultimately.

To quote Kuhn (1962: 172) once more, the analogy that relates the evolution of organisms to the evolution of scientific models is "nearly perfect." The testing of scientific theories through probabilistic verification does resemble the course of natural selection. Naturally any scientific advance involves a certain amount of commitment, in which one always runs the risk of being wrong, and the theory proposed here is indeed very audacious. I am very conscious of that, but there are some aspects here that appear to be in my favor. For one thing the gatekeepers of archeological knowledge have, through their censorship designed to preserve a dogma, also blocked scientific access to the information they have suppressed. The sciences, in their questions to archeology on many matters that are consequential to them, have been receiving a great deal of tainted information. Once they become aware of that, they are likely to react swiftly. While it is true that the protagonists in paradigmatic disputes tend to talk through each other, in this case I am not appealing so much to the Pleistocene archeologists, but more to those in the sciences who are interested in these matters, for a great variety of reasons—as well as to anyone else fascinated by the general topic. This is a deliberate strategy on my part, because I know from experience and from history that the discipline of archeology, like any hegemony based on authority, can absorb a great deal of dissent without change, by systematically disregarding it. For instance, the idea of taphonomic logic (Bednarik 1994), introduced in the early 1990s, has been ignored ever since because of its devastating effects on the credibility of most archeological claims. The discipline seems to be immune to such critique, but it depends very heavily on the approval of the public and other disciplines. After all, the public, society as a whole, underwrites the cost of the hobby of archeologists, who produce nothing of economic worth but have become a threat to the cultural heritage of the world. Not only do they destroy archeological deposits by excavating them, some of them facilitate the systematic destruction of rock art sites for payment (Bednarik 2008e). The dark sides of a primarily political discipline appropriating the interpretation of other societies' histories for the "winners of history" (today's nation states) are unknown to the public; archeology has always been careful to ingratiate itself with its sponsor. The demand that Pleistocene archeology comes clean with its blind spots, therefore, needs to come from the public, and in the specific field of human origins, both the public and the scientists need to exact valid information in place of the falsities and half-truths they have been fed so far.

It is simply untrue that archeologists understand human origins; it is even untrue that the dominant model presented to the public is the best available hypothesis (Bednarik 1996); and it is a perfectly reasonable entitlement of the public to expect not to be presented with some ridiculous Biblical story of an African Eve and her genocidal prodigy, reminding us of our collective guilt. If this is the best Pleistocene archeology can provide to explain the process of becoming human, then it is time for the public to withdraw its support from a discipline hopelessly mired in ignorance. If scholars cannot tell us what the primate bones from Liang Bua in Flores represent, perhaps it is time for them to begin reflecting on this academic impotence. And in this context I suggest to my opponents in Eve's camp to think twice before they apply the standard "Boucher de Perthes treatment" to me or to my propositions. It did not work then; it has never worked in the long term.

References

Aiello, L. C. and P. Wheeler 1995. The expensive-tissue hypothesis: the brain and the digestive system in human and primate evolution. *Current Anthropology* 36: 199–221.

Aoki, K. and M. W. Feldman 1991. Recessive hereditary deafness, assortative mating, and persistence of a sign language. *Theoretical Population Biology* 39: 358–372.

Ashley-Montagu, M. F. 1960. *An introduction to physical anthropology*. Thomas, Springfield, IL.

Badcock, C. R. 1980. *The psychoanalysis of culture*. Basil Blackwell, Oxford.

Barkow, J. H. 2001. Universals and evolutionary psychology. In P. M. Hejl (ed), *Universals and constructivism*, pp. 126–138. Suhrkamp Verlag, Frankfurt.

Beaumont, P. and R. G. Bednarik 2010. Pleistocene rock art from Africa. Paper presented to IFRAO Congress 2010, Foix.

Bednarik, R. G. 1988. Comment on F. D. McCarthy, 'Rock art sequences: a matter for clarification'. *Rock Art Research* 5(1): 35–38.

Bednarik, R. G. 1989. The Galgenberg figurine from Krems, Austria. *Rock Art Research* 6(2): 118–125.

Bednarik, R. G. 1993. Palaeolithic art in India. *Man and Environment* 18(2): 33–40.

Bednarik, R. G. 1994. A taphonomy of palaeoart. *Antiquity* 68: 68–74.

Bednarik, R. G. 1995. Concept-mediated marking in the Lower Palaeolithic. *Current Anthropology* 36: 605–634.

Bednarik, R. G. 1996. Crisis in Palaeolithic art studies. *Anthropologie* 34(1): 123–130.

Bednarik, R. G. 1997. The role of Pleistocene beads in documenting hominid cognition. *Rock Art Research* 14: 27–43.

Bednarik, R. G. 2003. The earliest evidence of palaeoart. *Rock Art Research* 20(2): 89–135.

Bednarik, R. G. 2006. The Middle Palaeolithic engravings from Oldisleben, Germany. *Anthropologie* 44(2): 113–121.

Bednarik, R. G. 2007. Antiquity and authorship of the Chauvet rock art. *Rock Art Research* 24: 21–34.

Bednarik, R. G. 2008a. Cupules. *Rock Art Research* 25: 61–100.

Bednarik, R. G. 2008b. The mythical Moderns. *Journal of World Prehistory* 21(2): 85–102.

Bednarik, R. G. 2008c. Children as Pleistocene artists. *Rock Art Research* 25(2): 173–182.

Bednarik, R. G. 2008d. The domestication of humans. *Anthropologie* 46(1): 1–17.

Bednarik, R. G. 2008e. More on rock art removal. *South African Archaeological Bulletin* 63(187): 82–84.

Bednarik, R. G., G. Kumar, A. Watchman and R. G. Roberts 2005. Preliminary results of the EIP Project. *Rock Art Research* 22: 147–197.

Bickerton, D. 2010. *Adams tongue: how humans made language, how language made humans*. Hill and Wang, New York, NY.

Biesele, M. 1993. *Women like meat. The folklore and foraging ideology of the Kalahari Ju/'Hoan*. Witwatersrand University Press, Johannesburg.

Bloom, H. 2001. *Global brain: the evolution of mass mind from the big bang to the 21st century*. Wiley, New York, NY.

Boehm, C. 1993. Egalitarian society and reverse dominance hierarchy. *Current Anthropology* 34: 227–254.

Boehm, C. 1999. *Hierarchy of the forest: the evolution of egalitarian behavior*. Harvard University Press, Cambridge, MA.

Boyd, R. and P. J. Richerson 2005. *The origin and evolution of cultures*. Oxford University Press, New York, NY.

Brace, C. L. 1996. Racialism and racist agendas: race, evolution, and behavior: a life history perspective. J. Philippe Rushton. *American Anthropologist* 98(1): 176–177.

Brace, C. L. 2005. *'Race' is a four-letter word: the genesis of the concept*. Oxford University Press, New York, NY.

Buss, D. M., M. Abbott, A. Angleitner, A. Biaggio, A. Blanco-Villasenor, M. Bruchon-Schweitzer et al. 1990. International preferences in selecting mates: a study of 37 societies. *Journal of Cross Cultural Psychology* 21: 5–47.

Buss, D. M. and M. L. Barnes 1986. Preferences in human mate selection. *Journal of Personality and Social Psychology* 50: 559–570.

Carden, N. 2009. Prints on the rocks: a study of the track representations from Piedra Museo locality (southern Patagonia). *Rock Art Research* 26: 29–42.

Cavalli-Sforza, L. L. and M. W. Feldman 1973. Cultural vs. biological inheritance. *American Journal of Human Genetics* 25: 618–637.

Conard, N. J. 2009. A female figurine from the basal Aurignacian of Hohle Fels Cave in southwestern Germany. *Nature* 459: 248–252.

Deacon, T. 1997. *The symbolic species. The co-evolution of language and the human brain*. Penguin Books, London.

De Beer, G. R. 1930. *Embryology and evolution*. Oxford University Press, Oxford.

De Beer, G. R. 1940. *Embryos and ancestors*. Oxford University Press, Oxford.

Delluc, B. and G. Delluc 1978. Les manifestations graphiques aurignaciens sur support rocheux des environs des Eyzies (Dordogne). *Gallia Préhistoire* 21: 213–438.

Dobzhansky, T. 1962. *Mankind evolving: the evolution of the human species*. Yale University Press, New Haven, CT.

Dobzhansky, T. 1972. On the evolutionary uniqueness of man. In T. Dobzhansky, M. K. Hecht and W. C. Steere (eds), *Evolutionary biology*, pp. 415–430. Appleton-Century-Crofts, New York, NY.

Dunbar, R. 2003. The social brain: mind, language, and society in evolutionary perspective. *Annual Review of Anthropology* 32: 163–181.

Durham, W. H. 1991. *Coevolution: genes, culture, and human diversity*. Stanford University Press, Stanford, CA.

Fedele, F. G. and B. Giaccio 2007. Paleolithic cultural change in western Eurasia across the 40,000 BP timeline: continuities and environmental forcing. In P. Chenna Reddy (ed), *Exploring the mind of ancient man. Festschrift to Robert G. Bednarik*, pp. 292–316. Research India Press, New Delhi.

Feldman, M. W. and L. L. Cavalli-Sforza 1989. On the theory of evolution under genetic and cultural transmission with application to the lactose absorption problem. In M. W. Feldman (ed), *Mathematical evolutionary theory*, pp. 145–173. Princeton University Press, Princeton, NJ.

Fischer, E. 1914. Die Rassenmerkmale des Menschen als Domestikationerscheinungen. *Zeitschrift für Morphologie und Anthropologie* 18: 479–524.

Francis, P. 1989. The manufacture of beads from shell. In C. F. Hayes III (ed), *Proceedings of the 1986 Shell Bead Conference: selected papers*, pp. 25–36. Research Records 20, Rochester Museum and Science Center, Rochester.

Fuentes, A. 2009. *Evolution of human behavior*. Oxford University Press, New York, NY/Oxford.

Gebser, J. 1985. *The ever-present origin*. Ohio University Press, Athens, OH.

Grammer, K. and R. Thornhill 1994. Human facial attractiveness and sexual selection: the role of symmetry and averageness. *Journal of Comparative Psychology* 108: 233–242.

Haldane, J. B. S. 1932. *The causes of evolution*. Longmans, Green & Co., London, and Harper Brothers, New York, NY.

Henneberg, M. 1990. Brain size/body weight variability in *Homo sapiens*: consequences for interpreting hominid evolution. *Homo* 39: 121–130.

Hodgson, D. 2000. Art, perception and information processing: an evolutionary perspective. *Rock Art Research* 17: 3–34.

Hodgson, D. 2003. Seeing the 'unseen': fragmented clues and the implicit in Palaeolithic art. *Cambridge Archaeological Journal* 13: 97–106.

Jablonka, E. and M. Lamb 2005. *Evolution in four dimensions: genetic, epigenetic, behavioral, and symbolic variation in the history of life*. MIT Press, Cambridge, MA.

Jones, D. M. 1995. Sexual selection, physical attractiveness and facial neoteny: cross-cultural evidence and implications. *Current Anthropology* 36(5): 723–748.

Jones, D. M. 1996. An evolutionary perspective on physical attractiveness. *Evolutionary Anthropology* 5(3): 97–109.

Klein, R. G. and B. Edgar 2002. *The dawn of human culture: a bold new theory on what sparked the 'big bang' of human consciousness*. Wiley, New York, NY.

Kuhn, T. 1962. *The structure of scientific revolutions*. University of Chicago Press, Chicago, IL.

Laland, K. N. 1994. Sexual selection with a culturally transmitted mating preference. *Theoretical Population Biology* 45: 1–15.

Laland, K., J. Odling-Smee and M. W. Feldman 2000. Niche construction, biological evolution and cultural change. *Behavioral and Brain Sciences* 23: 131–175.

Leonard, W. R. 2002. Food for thought: dietary change was a driving force in human evolution. *Scientific American* 287(6): 106–115.

Leonard, W. R. and M. L. Robertson 1992. Nutritional requirements and human evolution: a bioenergetics model. *American Journal of Human Biology* 4: 179–195.

Leonard, W. R. and M. L. Robertson 1994. Evolutionary perspectives on human nutrition: the influence of brain and body size on diet and metabolism. *American Journal of Human Biology* 6: 77–88.

Leonard, W. R. and M. L. Robertson 1997. Comparative primate energetics and hominid evolution. *American Journal of Physical Anthropology* 102(2): 265–281.

Lock, A. and C. R. Peters 1996. *Handbook of human symbolic evolution*. Clarendon Press, Oxford.

Malotki, E. 2010. Precursors of figurative depictions in the noniconic 'Western Archaic Tradition' rock art of the American West. *Pré-actes du congrès IFRAO Ariège 2010,* Pleistocene art of the Americas, DVD, Lacombe, Tarascon-sur-Ariège.

Mania, D. and U. Mania 1988. Deliberate engravings on bone artefacts of *Homo erectus*. *Rock Art Research* 5: 91–107.

Mellars, P. 2005. The impossible coincidence. A single-species model for the origins of modern human behavior in Europe. *Evolutionary Anthropology* 14: 12–27.

Mendel, J. G. 1866. Versuche über Pflanzen-Hybriden. *Verhandlungen des naturforschenden Vereines in Brünn*, 4 Abhandlungen, pp. 3–47.

Mesoudi, A., A. Whiten and K. Laland 2006. Towards a unified science of cultural evolution. *Behavioural and Brain Sciences* 29: 329–383.

Mithen, S. 1999. Symbolism and the supernatural. In R. Dunbar, C. Knight and C. Power (eds), *The evolution of culture: an interdisciplinary view*, pp. 147–172. Edinburgh University Press, Edinburgh.

Mithen, S. J. 2004. *After the Ice Age: a global human history 20,000-5000 BC*. Harvard University Press, Cambridge, MA.

Noble, W. and I. Davidson 1996. *Human evolution, language and mind: a psychological and archaeological enquiry*. Cambridge University Press, Cambridge.
Odling-Smee, F. J., K. N. Laland and M. W. Feldman 2003. *Niche construction: the neglected process in evolution*. Princeton University Press, Princeton, NJ.
Oyama, S. 2000. *Evolution's eye: a systems view of the biology-culture divide*. Duke University Press, Durham, NC.
Oyama, S., P. E. Griffiths and R. D. Gray 2001. *Cycles of contingency: developmental systems and evolution*. MIT Press, Cambridge, MA.
Porr, M. 2010. The Hohle Fels 'Venus': some remarks on animals, humans and metaphorical relationships in early Upper Palaeolithic art. *Rock Art Research* 27(2): 147–159.
Read, H. 1954. Art and the evolution of consciousness. *The Journal of Aesthetics and Art Criticism* 13(2): 143–155.
Richerson, P. J. and R. Boyd 2005. *Not by genes alone: how culture transformed human evolution*. University of Chicago Press, Chicago, IL.
Schwarcz, H. P., R. Grun, A. G. Lathham, D. Mania and K. Brunnacker 1988. The Bilzingsleben archaeological site: new dating evidence. *Archaeometry* 30: 5–17.
Sedikides, C., J. J. Skowronski and R. I. M. Dunbar 2006. When and why did the human self evolve? In M. Schaller, J. A. Simpson and D. T. Kenrick (eds), *Evolution and social psychology*, pp. 55–80. Psychology Press, New York, NY.
Shackelford, T. K. and R. J. Larsen 1997. Facial asymmetry as an indicator of psychological, emotional, and physiological distress. *Journal of Personality and Social Psychology* 72(1): 456–466.
Shiner, L. 2001. *The invention of art: a cultural history*. University of Chicago Press, Chicago, IL.
Silk, J. B. 2007. Social component of fitness in primate groups. *Science* 317: 1347–1351.
Steguweit, J. 1999. Intentionelle Schnittmarken auf Tierknochen von Bilzingsleben — Neue lasermikroskopische Untersuchungen. *Praehistoria Thuringica* 3: 64–79.
Steiner, R. 1964. *The arts and their mission, lectures 1923*. The Anthroposophic Press, New York, NY.
Steiner, R. 1970. *Man as symphony of the creative word, lectures 1923*. Rudolf Steiner Press, London.
Steiner, R. 1972. *A modern art of education, lectures 1923*. Rudolf Steiner Press, London.
Steiner, R. 1990. *Toward imagination: culture and the individual*. Anthroposophic Press, New York, NY.
Terberger, T. 1998. Endmesolithische Funde von Drigge, Lkr. Rügen — Kannibalen auf Rügen? *Jahrbuch für Bodendenkmalpflege Mecklenburg-Vorpommern* 46: 7–44.
White, R. 1995. Comment on R. G. Bednarik, 'Concept-mediated markings in the Lower Palaeolithic'. *Current Anthropology* 36: 623–625.
White, R. 2003. *Prehistoric art: the symbolic journey of humankind*. Harry N. Abrams, New York, NY.
Wild, E. M., M. Teschler-Nicola, W. Kutschera, P. Steier, E. Trinkaus and W. Wanek 2005. Direct dating of Early Upper Palaeolithic human remains from Mladeč. *Nature* 435: 332–335.

Chapter 6
The Big Picture

A Preamble

Some of the previous chapters would tend to give the impression that this book focuses on the origins of what some define as anatomically modern humans. This is attributable to the need to place the general topic of *the human condition* within a framework of explaining why dominant paradigms of recent decades are false, both in the general sense (i.e., why the dominant epistemology can only yield false models) and in the specific sense (i.e., why notions of the beginnings of human modernity fail). In that sense, this particular debate was crucial, having shown the inherent deficiencies in the discipline; why they are predictable; and how the issues relating to the final Pleistocene percolate right down and affect all preceding chapters of human development in similar patterns. We have seen that comprehensive evidence of symbol use extends back not, as the orthodox model demands, 30 or 40 ka, but at least twenty times as long. This is not only in the form of objects and markings collectively defined as paleoart, or in the evidence of pigment use. One form of symboling, i.e., complex language use, is demonstrated by the ability of maritime colonization, which extends in the order of one million years into the past. It is simply impossible for a group large enough to found a genetically viable population to embark on such an expedition—and the planning and preparation it engenders—without a minimum level in the complexity of communication. Whatever that level is (and one can design testable propositions about it), it has been demonstrated for the final part of the Early Pleistocene, through the several successful ocean crossings by colonizing parties that have occurred.

Human modernity, then, did not suddenly appear at any particular point in time, but developed gradually, over many hundreds of millennia. But human development as such began earlier, prior to the skilled use of symbols. Hominins first evolved during the Pliocene period, which lasted from 5.2 to 1.7 million years (Ma) ago. Earlier contenders such as *Sahelanthropus tchadensis* are even 7 Ma old, thought to have walked upright, but the Pliocene is the period during which the australopithecines flourished, apparently only in Africa. They occurred in several forms, at least one of which is thought by some to be a human ancestor. The earlier *Orrorin tugenensis* from Kenya, at about 6 Ma, is another contender for human ancestry. The position

R.G. Bednarik, *The Human Condition*, Developments in Primatology: Progress and Prospects, DOI 10.1007/978-1-4419-9353-3_6, © Springer Science+Business Media, LLC 2011

of the two *Ardipithecus* "species," *Ardipithecus ramidus* and *Ardipithecus kadabba*, remains controversial. They lived about 4.4 Ma ago.

The gracile australopithecines commence around 4.2 Ma years ago. With a brain little more than a third the size of a modern human's, they were certainly bipedal, as especially the Laetoli tracks amply demonstrate. Although they possessed a hyoid bone (Alemseged et al. 2006), it seems unlikely that they were capable of complex verbal communication. The Makapansgat cobble we visited in Chapter 3, found among their remains in the filling of a dolomite cave in South Africa, may perhaps not have been deposited by them, but by a contemporary human resembling *Kenyanthropus platyops*. The australopithecines are represented by several "species," *Australopithecus anamensis* (4.2–3.9 Ma), *A. afarensis* (3.6–2.9 Ma), *A. africanus* (3–2 Ma), *A. bahrelghazali*, and *A. garhi* (c. 2.5 Ma). The latter has been found together with stone tools and butchered animal remains, and has been suggested to link australopithecines to the later hominin lineage.

In view of the common use of tools by modern chimpanzees and other primates there can be no doubt that tool use increased during the long reign of the australopithecines. A particularly interesting aspect of the gracile australopithecines is that they apparently evolved into robust forms, now subsumed under the genus *Paranthropus*. These were more muscular and their skeletal remains are somewhat larger and more robust, and they developed alongside human species, beginning about two million years ago, in the very last part of the Pliocene. There is again much disagreement concerning the status of the genus; some researchers see it as too primitive to compete with the contemporary *Homo* species, but it has also been shown to have used Oldowan-type stone tools, and one species, *Paranthropus robustus*, is credited with using both "advanced" tools and fire at Swartkrans, South Africa. Interestingly, this has been explained away as being evidence of imitation of human behavior, an unlikely explanation that we have already encountered toward the end of the human ascent ("Neanderthals" imitating the mythical "Moderns"). The currently earliest solid evidence of fire use is from the same country, excavated in the lowest horizons of Wonderwerk Cave (Fig. 6.1); and about 1.7 Ma old (Beaumont in press). The two other *Paranthropus* "species" currently distinguished are *P. boisei* (2.3–1.4 Ma) and *P. aethiopicus* (2.7–2.3 Ma).

Kenyanthropus platyops lived about 3.5–3.3 Ma years ago and has been proposed by some to be human, perhaps ancestral to *Homo habilis*. Others include *K. platyops* with the australopithecines, suggesting that the species is not distinctive enough to warrant having its own genus. On the other hand, some prefer to group *Homo rudolfensis* with this genus, calling it *Kenyanthropus rudolfensis*.

The fully human species that existed alongside australopithecines and then later coincided with *Paranthropus* were *Homo habilis*, *H. rudolfensis*, and *H. ergaster*. *H. rudolfensis* dates from about 2.5 to 1.9 Ma ago, and the roughly contemporary *H. habilis* lived 2.3–1.6 Ma ago. Numerous stone tools of the Oldowan tradition have been found with the latter's remains, and were probably also used by the former. Some researchers consider them a single species, but it is more widely thought that they were too different. There is much unresolved speculation about which of the two, if either, was the ancestor of the subsequent hominins, but *H. rudolfensis* seems

Fig. 6.1 Part of the Oldowan hearth in Wonderwerk Cave, an ash lens of 2 m length, containing hundreds of calcined bone fragments, c. 1.7 million years old

more developed. The *H. habilis* brain of 650 ml was considerably larger than the average *Australopithecus* brain, and the brain shape was more human-like. In one of its endocasts, the bulge of Broca's area, related to the motor control of speech, is visible, suggesting that the species may have been capable of verbal communication. The brain size of *H. rudolfensis*, with an average cranial capacity of 750 ml (the modern average is 1350 ml, ranging from 900 to 1880 ml, that of australopithecines ranges from 400 to 545 ml), considerably exceeding that of *H. habilis*. *H. rudolfensis* brains show a pattern that may be related to handedness and tool-use or manufacture. The degree of cranial asymmetry also appears to increase in later hominins. The endocast of the type specimen of *H. rudolfensis*, KNM-ER 1470, shows a somewhat more complex and modern-human-like third inferior frontal convolution compared with those of pongids. This is one of the few sources of information we have of the evolutionary reorganization of the brain, because the area it concerns includes Broca's area. Unfortunately, later hominin endocasts, from *H. habilis* and *H. erectus* through archaic *H. sapiens* to the present, seldom show the sulcal and gyral patterns faithfully. With regard to brain reorganization, left-right cerebral hemispheric asymmetries exist in extant pongids and the australopithecines, but neither the pattern nor direction is as strongly developed as in modern or fossil *Homo* (see next chapter).

Sexual dimorphism, the strong differences of the sexes apparent in the early hominins, reduced gradually with the australopithecines, and by the time of *Homo ergaster* (848–908 ml) had largely disappeared. This species, or subspecies of early *H. erectus*, existed between 1.9 and 1.4 Ma ago and it heralds significant changes, such as the production of typologically distinctive stone tools, most importantly "handaxes." It is also the first that can be shown to indisputably having left Africa, although the matter of possible previous human colonizers remains open. In particular, several human remains from Dmanisi, Georgia, 1.7 Ma old and with a brain ranging from 600 to 780 ml (Gabunia and Vekua 1995; Dean and Delson

1995), seem to be closer to *H. habilis* than to *H. ergaster*. More importantly, the stone tools at Dmanisi are of the Oldowan cobble tool type, not of the early Acheulian associated with *H. ergaster*. However, the earliest supposedly hominin finds reported from Asia are the late Pliocene mandibular fragment with two teeth and a single maxillary incisor from the Longgupo Cave site in China (Huang and Fang 1991; Huang et al. 1995; Wood and Turner 1995). Paleomagnetic dating suggests an age of 1.96–1.78 million years ago, electron spin resonance dating has provided a conservative minimum age of a million years, and this is supported by the plentiful accompanying faunal remains (116 species). Together with the very early dates for Javan *Homo erectus*, this find questions the sole African development of *Homo* (Culotta 1995). Lacking good diagnostic features, the Longgupo specimens were thought to resemble either *Homo habilis* (Ciochon 1995) or *Homo ergaster* (Wood and Turner 1995), but alternative possibilities offer themselves. Two stone tools were found in the hominid/hominin-bearing sediment stratum of the cave. While the upper lateral incisor is generally accepted as being human, the mandibular fragment is not, being attributed to an ape by some commentators (Schwartz and Tattersall 1996). Dennis A. Etler and Milford H. Wolpoff (personal communication Nov. 1996) have both expressed the opinion that the latter might be of *Lufengpithecus*, but the incisor resembles both recent Asian human specimens and *H. erectus*. This would be as difficult to reconcile with the reported age as would be the occurrence of stone tools with an ape. However, it would be only "slightly" older than the claimed earliest date of that species in Java (Swisher et al. 1994) and the hominins at Dmanisi. Moreover, in 2000, stone tools found in Renzi Cave, Anhui Province, eastern China, were assigned a date of 2.25 million years, fuelling debate about the origins of *H. erectus*, a species widely but not universally accepted as ancestral to modern humans. The very early Chinese finds derive much support from those at the Pakistani sites Riwat and Pabbi Hills, dated to the Plio-Pleistocene and the Early Pleistocene (Rendell et al. 1989; Hurcombe 2004).

The Human Ascent

It then seems that the somatic conditions of producing the kinds of differentiated sounds that form speech were present in hominins by more than 3 Ma ago, and that by 2 Ma ago, some form of incipient speech was probably in use. Seen within the greater scheme of things, this would appear to be a realistic interpretation of the evidence currently available. By 1.7 Ma, at the latest, controlled use of fire is well illustrated. The hearth in Wonderwerk Cave is a good 30 m into the cave; it is about 2 m long and contains hundreds of smashed and calcined animal bones (Fig. 6.2). In other words, by the time *H. erectus* emerged and colonized huge parts of the Old World, he was very probably assisted in this by symboling competence and fire use. The notion that he arose in one particular place (Africa being favored by most) is perhaps simplistic: if people 1.8 Ma ago occupied large parts of Africa and Asia, as appears to be the case, it is more likely that some populations were contiguous,

Fig. 6.2 Detail of the oldest currently known hearth, Wonderwork Cave, South Africa, showing calcined bone fragment in the ash layer

and that reticulate gene flow occurred—which means that the newer forms did not evolve in one place, but as a reflection of developments elsewhere. This is simply the same situation that we encounter toward the end of the Pleistocene, and very probably throughout the intervening times.

Homo erectus first appears 1.8 Ma ago with a brain capacity of about 850 ml, but its late Asian representatives, in the order of 300–500 ka old, ranged from 1100 to 1250 ml, i.e., well within the range of modern humans. For instance, the cranial capacity of present-day female Aborigines is 1181 ml; that of males is 1347 ml. The increase occurred without any apparent change in body size, *H. erectus* is of modern height, stature, and body proportions. The species' fossils occur widely in Africa and Asia, but European claims of its occurrence are rejected by many scholars. In China, its remains were initially found at the Zhoukoudian 1 Site, later at Gongwangling, Chenjiawo, Donghecun, Qizianshan, Tangshan, Longtandong, Xiaohuashan, Xichuan, Danawu, Quyuan River Mouth, Yunxia, Longgudong-Yunxian, Longgudong-Jianshi, and Bailongdong. First found at Trinil on the Solo River (Dubois 1894), Early Pleistocene hominins were also recovered from the upper part of the Pucangan beds and the probably later lower part of the Kabuh beds at Sangiran, Java. Klaatsch (1908) was the first to propose an evolutionary sequence connecting the Javanese hominins with Australids. After the discovery of much more recent hominin remains at Ngandong, Weidenreich (1943, 1945, 1951) developed this idea (see Fig. 1.1), which was eventually incorporated in what is today known as the multiregional hypothesis of the origins of modern humans (Larnach and Macintosh 1974; Thorne 1980, 1989; Thorne and Wilson 1977; Thorne and Wolpoff 1981; Wolpoff 1980, 1989, 1991, 1997). The Mojokerto child is the earliest known hominin fossil in Java, being from the Pucangan deposits and 1.81 Ma old (Swisher et al. 1994), while the Sangiran individuals are about 1.66 Ma old.

Despite a considerable number of hominin fossils from the Middle Pleistocene, the subsequent evolutionary history of humans in Asia defining the grading of *H. erectus* into archaic *Homo sapiens* remains unclear. Chinese fossils of early archaic *H. sapiens* of erectoid features are those from Xujiayao, Dingcun, Yunxia, Yenshan and the Dali cranium, and Wanlongdong maxilla, which are already of the early Late Pleistocene, but still of very robust features. The picture becomes even more confusing when we consider the only two Indian finds, of the last part of the Middle Pleistocene and found with a rich assemblage of Acheulian tools. The Hathnora calotte from the Narmada valley was initially described as a late *H. erectus* (de Lumley and Sonakia 1985), and while its thick torus, postorbital constriction and bone thickness suggest this, the vault is far too well rounded and of exceptional size (Fig. 6.3). At approximately 1300 ml it is not only above the range of *H. erectus*, but also high for a *H. sapiens*, since it is thought to be of a female in her thirties. Only the right half of the cranium, with zygomatic arch, right torus and right part of occipital are intact (Bednarik 1995: 611, 1997: 26). The second Indian specimen, from the same site and stratum, but from another individual and perhaps even another species, is an adult clavicle that suggests a body size of little over a meter (Bednarik et al. 2005).

Combined with the recently discovered new hominin dwarf specimens from Flores, this serves to underline the extremely fragmentary nature of our evidence of hominin evolution in Asia. This is particularly unfortunate because the current cultural evidence suggests that some of the major developments in nonphysical human evolution occurred in southern Asia. Certainly, the hominin specimens in Africa

Fig. 6.3 The Hathnora calotte, Narmada valley, India, perhaps around 200,000 years old

and eastern Asia since the final Pliocene demand that southern Asia must have been occupied for at least two million years. The recent discovery of stratified Olduwan-type cobble tool industries, particularly in the basal layer of Daraki-Chattan, central India, adds a tantalizing note to this scenario (Bednarik et al. 2005). The most parsimonious interpretation of the data as it stands appears to be that pre-*H. erectus* hominins were established in both southern and eastern Asia.

But what is particularly perplexing is the presence of sound evidence of rock art production, as noted in Chapter 3, from two central Indian deposits of Oldowan-like Mode 1 traditions, first at Auditorium Cave at Bhimbetka, and most securely demonstrated at Daraki-Chattan, in the Chambal valley. Twenty-eight cupules on exfoliated rock fragments (and one on *in-situ* rock) were excavated from Acheulian Mode 2 layers as well as from the chopping tool industry below them. Moreover, numerous hammerstones used in the production of the 530 cupules in the quartzite cave were recovered from the Mode 1 technocomplex of the site's basal sediments. Although secure dating of the sequence has remained elusive so far at Daraki-Chattan, the "handaxe"-free industries in India are known from many other sites.

The earliest phase of human presence in India, of Mode 1 assemblages, is represented by limited but tantalizing references to archaic chopping tools, cores, and flake tools, sometimes referred to as Soanian. Most of these occurrences are surface finds (e.g., Guzder 1980; Sharma and Roy 1985; or Pabbi Hills in Pakistan, Hurcombe 2004) or come from alluvial or colluvial deposits, including conglomerate horizons (e.g., Durkadi, Armand 1983; or Mahadeo-Piparia, Khatri 1963). In very few cases, the Mode 1 industries have been excavated from secure stratigraphies, Auditorium Cave and Daraki-Chattan being the prime examples. There remains wide disagreement about the antiquity of the early Acheulian and the Mode 1 industries of India, reflecting similar recent debates in southern Europe (i.e., in Spain, France, and Italy). Based on the potassium-argon dating of volcanic ash in the Kukdi valley near Pune to 1.4 million years ago, some favor that magnitude of age for the earliest phase of the Acheulian (Misra and Rajaguru 1994; Badam and Rajaguru 1994). An age of well over 400 ka seems also assured by thorium-uranium dating (Mishra 1992; Misra and Rajaguru 1994). Others, especially Acharyya and Basu (1993), reject such a great antiquity for the early Acheulian in the subcontinent, and the ESR date of c. 1.2 Ma for early Acheulian finds at Isampur (Paddayya et al. 2002) remains tentative. However, Chauhan and Patnaik (2008) have shown that lithics at the Narmada site Dhansi, less than 3 km south of the hominin site Hathnora, occur in a major formation of the Matuyama Chron, presumably placing them in the Early Pleistocene.

Until the chronology of the early Lower Paleolithic traditions of southern Asia can be clarified, little should be said about the possible antiquity of the extremely early rock art of central India. However, combined with other factors, such as the first development of maritime navigation, also in southern Asia, there is a reasonable expectation that this region could have been a major hub in the Early Pleistocene development of human cognition, technology, and culture. Whereas the early physical evolution of hominins can reasonably be attributed to Africa, there is currently no evidence from that continent that would imply significant *cultural* development prior to mid-Middle Pleistocene times. The early appearance of pigment use (Hunsgi) and collection of quartz crystals (Singi Talav), in both cases in the Lower Acheulian (see Chapter 3), combined with the certainly very complex behavior evidenced in the early cupule tradition predating the Acheulian, suggests southern Asia's priority in the cognitive development of hominins.

When I presented the evidence for the Pleistocene seafaring abilities one of the commentators chided me for stating that *Homo erectus* was the greatest colonizer in the phylogenetic history of primates and also the greatest achiever in a cultural sense. "Is there anyone that would actually accept such a view", he asked (Rowland 2003). This implies that he had not appreciated the qualifying content of my statement. I had pointed out that the capacity to domesticate natural systems and energies; of complex communication; of symboling; and presumably of perceiving constructs of reality—i.e., traits defining modernity in humans—were first developed by *H. erectus*. And I noted "what *H. sapiens sapiens* added to this cultural capital is much less significant, seen in a proper historical perspective". I pointed out that once these capacities had been initiated, all subsequent developments were logical outcomes of what had already been set in motion. As an archeologist Rowland, I assume, was concerned about two aspects of my audacious statement: that it utterly contradicts archeological dogma; and that it rejects the notion of modern humans being the crown of creation. His first concern would be understandable, many archeologists share it, but it is irrelevant to anyone wishing to understand human origins. Of more interest would be the second aspect: the propensity of human groups to glorify the achievements of their kin, always at the expense of "The Others." I fear that "in a proper historical perspective", Neil Armstrong's small step in 1969 was just a small step for mankind, when compared to the incredible accomplishments of our predecessors several hundred millennia back. The calculated risks of a trained specialist, who would never undertake a journey into the unknown without a strong belief in a safe return, are no match for the courage of taking one's clan on a one-way trip to possible, even probable, oblivion. We have ethnographic evidence that the failure rate of Stone Age sea crossings must have been horrific: Tindale (1962) recorded an average death rate of 50%. Nor can we find an equivalent transformation in the way the human condition changed during *H. erectus'* reign at any subsequent time: it seems that once a specific set of circumstances had determined the direction of nonsomatic development of humans, it led to a spiraling process of elaboration of the minutiae, which, subject to external variables of changing climate and environment, was more or less an inevitable outcome.

Maritime colonization demonstrates a whole raft of human capacities, such as the successful operation of a collective consciousness (*sensu* Émile Durkheim); a well-developed theory of mind (ToM); the effective use of language to convey abstract concepts and refer to future conditions; a certain minimum level of social organization; the capacity of long-term forward planning; and of course a variety of technological prerequisites.

The merkwelten of earlier hominins may not be accessible to us, but there are quite a few aspects of Middle Pleistocene people's reference frames that might be accessible to testing. To cross the sea, even a single individual would need to engage in planning such an expedition. Since the straits separating islands cannot be crossed by mere drifting, because of their always strong and unpredictable transverse currents, humans were never able to cross them without some form of flotation device capable of being propelled. But individuals or small groups cannot found a viable new population; this requires dozens of people, and most especially a good number of females of childbearing age. In all ethnographic cases we know of, only the males took to the sea *habitually*, for fishing, whaling, sealing, seabirding, warfare, and so forth, because the group's females were too precious to risk. It is, therefore, not unreasonable to imagine that to plan a maritime colonization would have involved a great social and organizational challenge. And that is before we consider the need to plan the technical details: rafts of adequate sizes did presumably not just happen to be around, they had to be built. Details such as the transport of drinking water had to be attended to, and of having the tools to survive not only the journey, but also the challenges immediately after landing. There should be no doubt that many of these efforts, perhaps even most of them, either failed immediately, with the migrants forced out onto the open ocean to perish there (Bednarik 2003a: 61); or at a later time when they failed to establish a new population in the new land, for whatever reason. And yet, we know that there were numerous instances when Pleistocene navigators succeeded in founding permanent new colonies, sometimes after traveling for hundreds of kilometers across open sea. For such island populations to become archeologically visible, they had to be very well established and number in thousands. We know that they were of Lower and Middle Paleolithic technological traditions. This, more than any other factor one can present, provides a level of the full capabilities of hominins of Modes 1–3, respectively, and any alternative, minimizing archeological claims of their abilities have to be, I regret having to say, relegated into the realm of mythology.

Hermeneutics of these issues hinges on sound and representative archeological data as well as the inputs of evolutionary psychology, neuroscience, semiotic modeling and similar pursuits. But not only is the quality of mainstream archeology's information largely inadequate and tainted by dogma, there is very little substantive dialogue between archeology and the various sciences concerned. While the study of cognition has not kept pace with the study of the brain, the disparity is still much greater between the advanced state of today's neuroscience and the primitive, backward operation of Pleistocene archeology. Humanists are guided, if only subconsciously, by an inclination to maintain the special status of humans, which is being gradually eroded by the sciences, such as for instance ethology. Virtually all the newly discovered human singularities are located in areas associated with either complex social cognition (theory of mind) or language, which renders it inevitable to look for the origins of human modernity in those areas.

Each organism can only prove the existence of his or her own mind through introspection, and has no direct access to others' minds. The presumption that other cognizing organisms have a mind is termed a *theory of mind* (ToM): the ability to attribute mental states—beliefs, intents, desires, pretending, knowledge, etc.—to oneself and others and to understand that others have beliefs, desires, and intentions that are different from one's own. Although present in numerous species, at greatly differing levels, it has perhaps attracted most attention in the study of two groups, children and apes, and the level they conceive of mental activity in others, attribute intention to and predict the behavior of others.

Theory of Mind, Language, and Exograms

One of the most comprehensive and influential recent syntheses on the connection between consciousness and language is the work of cognitive neuroscientist and research psychologist Merlin Donald (1991, 2001). He opposes the trend he perceives to deny evolutionary significance to human consciousness and holds that enculturation "as a possible formative process in its own right" has been neglected. In his natural history of particularly human, multi-layered consciousness, he focuses on the dramatic changes from nonhuman to human primates and on the need to explain the generation of culture if we are to understand consciousness fully.

Essentially, Donald recycles the idea, expressed by others before him, of a few distinctive stages in the hominin ascent, which he defines as "episodic," "mimetic" (representation of knowledge through voluntary motor acts), "mythic" (spoken language), and "graphic representation." Conscious experience, he states, is held together by representation (Donald 1993: 785). He defines nonhuman primate memory, and hence cognition, as episodic, and perceives the first major change with the appearance of *Homo erectus*. That species is credited with mimesis, but not with speech-like language. *Homo sapiens*, according to Donald, developed first "mythic culture," and through it language, which contrasts with the view of linguists such as Dunbar or Bickerton who would credit social exchange with the rise of language. Fairservice (1975) had proposed three similar stages of cultural evolution for humans, enactive, ikonic, and symbolic, following even earlier such proposals.

One can readily agree with Donald that humans both produce and are produced by culture. He holds that culture directs human attention, determines what we learn, and mediates even our most private thoughts. Very few of our thoughts are truly unique; we are mostly symbol manipulators and rarely symbolic inventors. There is a great deal in his work that prompts introspection, but the tangible predictions his extended think-piece offers are in the end significantly flawed. Donald's inadequate understanding of the anthropological data (Brace 1993, 1996, 1999), together with his lack of knowledge of the empirical evidence for early symboling, his poor presentation of the relevant neurology, and his neglect of cognitive ethology (Cynx and Clark 1993) all render his model questionable. C. L. Brace (1993) particularly decries the "cavalier misuse of information available from anatomy, anthropology, and archaeology". It is true that, for instance, Donald accepts Lieberman's notions of Neanderthal vocal capabilities (Lieberman and Crelin 1971; Lieberman et al. 1972, 1992), although they had long been refuted (Falk 1975, 1987; Arensburg et al. 1989, 1990; cf. Laitman et al. 1992)—and today are absurdities, in view of the australopithecine hyoid (Alemseged et al. 2006) and the FOXP2 language gene on chromosome 7 from "Neanderthal" remains (Krause et al. 2007; cf. Enard et al. 2002; Zhang et al. 2002; Sanjuan et al. 2006). Donald's flawed view of Neanderthal's lack of language then prompts him to assume severe cognitive limitations, while his undocumented notion that "Neanderthals underwent a drastic, rapid extinction" has already been defined as "unsupported assertion based on a kind of current "folk-wisdom" that has to be relegated to the realm of pop-science", comparable to phrenology (Brace 1993) and has been opposed by me for decades. Donald's contention that the introduction of language would speed up the rate of cognitive evolution is analogous to claiming that the rate of mutation determines the rate of genetic evolution (Cynx and Clark 1993).

Of all these and other limitations, I regard Donald's neglect of empirical evidence for early symboling, i.e., the material we reviewed in Chapter 3, as the most consequential; I perceive no value in devising a just-so (untestable) explanation for symboling that completely ignores the relevant evidence. His "best-guess sequence of what actually has happened in the human case, based on as much relevant empirical data [he] could find" (Donald 1993: 782) is precisely that, and can in view of his inadequate empirical database only be a false guess. Presenting such a hypothesis in a non-falsifiable format merely amounts to stating an opinion. The obvious problem here is that, in deferring to testability, we are in matters relating to human evolution to a very large degree at the mercy of the quality of the empirical data. For instance, a statement such as the following has been plucked out of thin air, it has no basis in the data, and it is typical of Donald's many assertions: "The rate at which [later] oral language capacity generated phonological, semantic, and grammatical variations across the inhabited human world would have been completely stupefying to poor old *Homo erectus*." Perhaps this is so, but it is certainly not justified by any testable data, and contradicted by the available *sound* data.

It is unclear where Donald obtained many of his data, especially archeological and anthropological claims, but it seems justified to assume that he placed far too much reliance on the now discredited notions of the universal replacement of all Late Pleistocene robust humans by a unique African species. That model, based as we have seen on fakery, excessive promotion, and the discipline's gullibility, is now refuted, robbing Donald's thesis of a credible time framework. Consequently there are clear problems with all of his stages. For instance a typical mimetic activity would be iconic drawing; yet *H. erectus*, Donald's quintessential representative of that stage, has left us no surviving examples. On the other hand, that species has provided clear evidence of complex language skills, for instance through its maritime colonizations. If we interpreted the record literally, this would invert Donald's mimetic and linguistic stages. I am not advocating that, I am merely noting that either Donald's sequence is wrong, or the stages he identifies are. Since he created these without presenting tenable evidence it is not possible to test them, and any refuting evidence would presumably only result in tinkering with his just-so model.

Most of the empirical shortcomings of Donald's thesis were apparent at the time his first book was published. For instance, Bickerton had (1990) proposed an evolutionary model of language development, suggesting that *H. erectus* had an unstructured protolanguage, followed by the establishment of an autonomous syntax by archaic *H. sapiens*. I had written about engrams and what Donald later called exograms (Bednarik 1987), and about the cognitive development of hominins (1990a, b, cf. 1992a). Although Donald's first volume could not benefit from the first attempt to create a comprehensive register of all very early material finds that might indicate symboling abilities (Bednarik 1992b), he did not avail himself of this rich mother lode in his 2001 volume. Today it is much more likely than not that habilines and *rudolfensis* possessed some form of speech, and that *erectus* had at their disposal somewhat developed and probably recursive verbal communication (for realistic scenarios, cf. Bickerton 1993, 1996, and especially 2010; Dunbar 1996; Aitchison 1996; Falk 2009). It has been claimed that all modern human languages are recursive (but see Everett 2005), and although nonrecursive language would be possible, chimpanzees have neither (Premack 2007). Notwithstanding Baron-Cohen's (1991) speculation that the inclination to spontaneously reference an object by "proto-declarative pointing" may be the underlying motive behind all human communication, and other endeavors to clarify language origins, this issue is far from any resolution. But it seems safe to conclude that hominins paid a very high biological and social price for their encephalization, so there has to have been a strong reason for such an adaptation to be selected for. Moreover, as Bickerton notes, the benefits of language use must have been immediately apparent, or else it would have not been selected for. Not only has there been a phenomenal increase in cranial capacity, there have also been, as we shall soon see, very significant changes in the actual brain tissue, between apes and present humans (see next chapter). The former process is well quantified through time, while we lack a time line for the latter. Nevertheless, it stands to reason that the earliest humans already used their brains in ways that differed considerably from those of apes. The demonstrated changes are not adequately explained by increasing tool use or social complexity. If social complexity of hominin society were to be the primary explanation for human encephalization, how could we account for the numerous instances of very complex social behavior in the animal kingdom, extending even to various invertebrates? It is much more judicious and realistic to implicate improved conscious experience and symboling, including language.

All this renders Donald's evolutionary model highly implausible, and in some respects refuted. Of much greater relevance is his development of the notion of external memory storage. As we noted in Chapter 4, this is again a recapitulation of much earlier work, but it has prompted a very productive debate over the following two decades. Some of the most interesting developments are "extended mind studies" and the preliminary conceptualizations of exograms. Donald's initial perception of the storage of symbolic information outside the brain followed Gregory's (1970) lead: "the exact external analog of internal, or biological memory, namely, a storage and retrieval system that allows humans to accumulate experience and knowledge" (Donald 1991: 309). This is an oversimplified rationalization, because obviously the external storage device is not analogous to its internal memory trace, and there are in fact significant differences between engrams and exograms.

With his "exograms" Donald (1991: 308–333, 2001: 305–315) has created another academic cottage industry in cognitive science. Exograms are the single entries in an external symbolic storage system, and Donald derived the term from "engram," as used by Lashley (1950) but previously coined by Semon (1921: 24), who also referred to it as "mnemic trace" (72 years before Dawkins' "meme"; Semon 1904). An engram (Bednarik 1987) is a hypothetical (but not so far demonstrated) single entry in a biological memory system (stored in response to external stimuli as a biophysical or biochemical change in neural tissue). Dawkins (1976) believes he rediscovered the concept independently. An exogram, then, would be an external memory record of an idea. Most importantly for our purposes, exograms are *made*, symbolic inventions that may have undergone a process of repeated examination, testing, and improvement. As Donald (1991: table 8.1) observes, engrams would be impermanent, of constrained format, fixed physical medium, limited capacity and size, and not easily refinable. In these and other characteristics they would differ fundamentally from exograms, which are often permanent, are unconstrained and reformatable, can be of any medium, have virtually unlimited capacity and size, and can be subjected to unlimited iterative refinement.

Since Donald noted in 1991 that "unlike the constantly-moving contents of biological working memory, the products of thinking, when reformatted exogrammatically, could be frozen in time, held up to scrutiny at some future date, altered and re-entered into storage", there has been much debate about the parity in stressing various dimensions of difference between exograms and engrams. For instance Adams and Aizawa (2001: 58) recommend that Donald should reject "extended mind" and embrace their conclusion that "there can be no cognitive science of transcorporeal processes" (see also Adams and Aizawa 2008; Aizawa and Adams 2005; Block 2005; Prinz 2006; Malafouris 2004; Rupert 2004). They demand that one must subscribe to either the hypothesis of *extended cognition* or the hypothesis of *extended cognitive systems*. If the system that accomplishes cognition comprises the brain together with extra-cranial aids, it needs to be justified why parts of it should be considered constituents of a cognitive process rather than causal contributors to cognitive processes taking place in the brain. Does the cognition of the blind man end at the point of his stick? Is the retina part of a cognitive system? If cognition were defined as just information processing, it would take place in, for instance, computers, among other external devices. But if cognition requires consciousness, external information processors are not cognizing, and "lower" organisms that can solve problems are either merely responding to stimuli or have also cognition. For Adams and Aizawa, cognition is defined by two factors: it is nonderived, intrinsic, or original content, whereas a computer's content, for instance, is derived. Secondly, for them cognitive processes are of a special kind, the mechanisms by which organisms remember, perceive, attend, and learn. Cognitive psychologists rely in the study of these phenomena on specific experimental devices, including reaction time experiments, recall tasks, and so forth. But this, to me, sounds like placing the cart before the horse: do the practices of a discipline determine the nature of the phenomenon, or should it not be the other way around? Although Adams and Aizawa cite three "theories of content" (Dretske 1981; Fodor 1990; Cummins 1996), explaining how original content arises naturally, none of them is universally accepted today. If we endorsed Menary's

(2007: 15) definition of cognition as the completion of a cognitive task by the manipulation of representations, we would need to know which tasks count as cognitive. Thus, agreement on a theory of cognition remains profoundly elusive.

This suggests perhaps that the "reductionist hardliners" often mentioned by Donald (2001) might be in a better position to bring some clarity into the issue. Consider, for instance, the operation of short-term memory, initially documented by Miller's (1956) seminal work. The finding that its capacity, about two seconds of sound, differs according to the language used (Baddeley 1992, 2000) links memory to language: Miller's 7±2 digits in English correspond to around six in Welsh, or ten in Chinese. The linear relationship between response time and correct response probability for word recall and recognition determined by Tarnow's most recent work (2009) suggests that the tagging level is the level of depletion of the readily releasable pool (RRP) of neurotransmitter vesicles at presynaptic terminals. The depletion of RRP (exocytosis) and short-term memory decay may be the ensuing recycling of the neurotransmitter vesicles (endocytosis), and the pattern of depleted presynaptic terminals corresponds to the long-term memory trace.

Such details are often disregarded by cognitive psychologists and others from the social sciences, or relegated to reductionism. This is related to the progressive diminution of the influence of the humanist disciplines. But in the interest of wresting control of such debates as that of the origins of human modernity from disciplines that foster only vague and poorly testable constructs, it must be said that the hard sciences are inevitably better equipped, provided that mindless reductionism is kept at bay. I will illustrate the point in the next chapter with specific examples, showing how hard data can point us in unexpected directions and greatly enrich the discussion of early human cognition, and of the condition of being human generally.

As Jacques and Zelazo (2005: 149) observe, "symbol use leads to increased psychological distance between the symbol user and the environment". Linguistic symbols, they note, are particularly well suited for creating such distance, because of their arbitrariness. On the one hand, labeling increases cognitive flexibility via the effect on level of consciousness; on the other the arbitrary nature of language allows for flexible thinking through distancing. Discursive thinking (which is conducted in linguistic terms; but note Bickerton's objection, 2010: 240–243) is intrinsically more flexible than thinking that is conducted on the basis of perceptual information, regardless of one's level of consciousness. There is a good deal of supporting evidence and literature to credit language with a key role in theory of mind performance, but there is also a case for this being a one-way relationship. Certainly there is strong evidence that language proficiency facilitates ToM formation; there is little to show that ToM has such effects on language ability.

In the final analysis there is nothing revolutionary about the idea of external storage of symbolic information. It is obvious that today's cultures exist almost entirely outside of our brains, precisely because, as Donald states, exograms have virtually unlimited capacity and size. Today they constitute almost all that is culture, and the modern individual is hopelessly overwhelmed by them, able to connect with only a minute part of them. The immense success of external, media-enabled human "cognition extensions" cannot be, however, explained without considering the properties of the media themselves and the strategies humans have developed to harness them. This is where Donald's thesis failed: what are the early forms of material evidence suggestive of mnemonic devices, or other objects of inherent but nonutilitarian meaning? Donald ignored this corpus of evidence, the exograms we need to consider if we are to chart the timing and the course of the developments they denote. We have considered this evidence in Chapter 3.

Ontogenic Plasticity and Neoteny

As hominins gained increased competence in representation and other cognitive and mental skills they forfeited neurological robusticity and apparently became susceptible to neurodegenerative diseases (see Chapter 7). It seems that every evolutionary improvement also involves some detriment. For instance, genetic bottlenecks may promote accelerated evolution through punctuated equilibriums, but they also tend to reduce genetic fitness in the population concerned (Bryant et al. 1986; Berger et al. 2008). Similarly, the neoteny of final Pleistocene and Holocene humans has been most detrimental to various of their physical aspects, but it has also involved highly advantageous changes, especially through the retention of "evolvability" or ontogenic plasticity (De Beer 1930)—which is emerging here as a key factor in the nonphysical evolution of humans.

One of the most promising perspectives the new hypothesis of human self-domestication offers on the question of the origins of paleoart and other forms of early exograms derives from this retention of plasticity. Biologically, the neoteny of our extant subspecies should hardly be a contentious issue, although there has been remarkably little consideration of it in the venerable discipline of paleoanthropology. Perhaps seeing ourselves as the noble creatures destined for great things, created in the image of a deity, has been preferred to the concept of a neotenous ape whose success is simply attributable to youthful flexibility, playful behavior, and infantile inquisitiveness. Throughout recorded human history, humans have shown a distinctive propensity for constructing self-flattering images of themselves, in every possible sense, so this is hardly surprising. This template is, however, not very conducive to scientific inquiry, and if the objective were a legitimate understanding of the processes of humanization, we would need less self-aggrandizement and more academic probity.

A classical example is the famous Franco-Cantabrian cave art of France and Spain, with its Upper Paleolithic artistic masterworks. These have been subjected to over a century of scholarly attention, after they were first virulently rejected by archeology for some decades. The result has been a long series of claims extolling the gravity, sophistication, and importance of this tradition, which is seen as marking the advent of proper culture and proving the obvious—its glorious invention in Europe. This laudable work has also resulted in many etic speculations about suitably profound meanings. There have been claims that the cave art is the work of shamans or has other deeply religious meaning, that it indicates hunting magic, totemism, animism, and so forth. The notion that much if not most of this paleoart is the work of children and teenagers has attracted rather less enthusiasm. Indeed, to the traditional scholar of Franco-Cantabrian "art" it may sound almost sacrilegious, and yet the hard evidence suggests its cogency. There is in fact very little evidence that any major portion of this celebrated corpus is the work of adults, but there is sound evidence that juveniles created much of it (Bednarik 1986, 2002, 2008; Guthrie 2005). Although most rock art or other paleoart cannot be securely attributed to specific age groups, there are some types of art-like remains that present adequate forensic evidence to permit such attribution securely. Most important and unambiguous among these, in the context of European Pleistocene paleoart, are (1) finger flutings; (2) prints and stencils of body parts; and (3) the fingertip stamp marks made with paint on certain portable objects. There are other situational conditions (e.g., aperture size of only available access to a site, or widths of wet-applied pigment lines drawn with fingers) that may permit limited deductions concerning body size, but they are less persuasive. However, there is also a good deal of information available, concerning the age of cave visitors, from impressions of feet, hands, and other body parts, on clay floors and on soft wall deposits (Bednarik 2008).

Quantified forensic evidence shows that the finger flutings of the caves of Europe and Australia, most of which are of the Pleistocene, are overwhelmingly made by children or teenagers (Bednarik 1986; Sharpe and Van Gelder 2006). Prints and stencils of body parts, notably hands, in the Franco-Cantabrian traditions are entirely the work of young people (Guthrie 2005). The finger stamp marks commonly found on some Magdalenian plaques were also made by children, perhaps six to ten years old (Bednarik 2002). Finally, the overwhelming majority of the hundreds of human tracks found in no fewer than eleven Paleolithic "art" caves of Europe, certainly well over 90% of them, are by juveniles, some as young as three, most falling between the ages of nine to fifteen (Clottes 1985, 1986, 1997: 31; Clottes and Courtin 1995: 175; Duday and Garcia 1983, 1985, 1990; Garcia 2003; Garcia and Duday 1993; Pales 1954, 1960, 1976; Roveland 2000).

Therefore, the available record indicates a distinct bias in favor of children's markings, among those types of surviving paleoart that permit reliable determination, mirrored in the ages indicated by the surviving human tracks. While it is obvious that none of the footprints on cave floors need to necessarily relate to any of the cave art of such sites, it is equally obvious that there would be expected to be a much greater number of adult footprints if adults had significantly contributed to the "art." It appears, therefore, extremely unlikely that the pattern is merely a sampling phenomenon. Unless we were to postulate that *only* those forms of paleoart permitting age estimates of the artists were for some cultural reason made by children and adolescents—which logically seems to be beyond reasonable probability—we need to accept that there is a very high probability that other paleoart forms were also often the work of young people. This would be supported by the sizes of footprints observed on cave floors. The alternative hypothesis, that all or most other Pleistocene paleoart in Europe is the preserve of adults has no empirical support.

This is not intended to exclude the possibility that adults *did* create a certain portion of European Pleistocene rock art; there is no proof that it is exclusively the work of young people. But what careful consideration of the empirical evidence does show is consistent with the observation that most of the explanatory endeavors offered for this famous European corpus since the late nineteenth century have in recent years been rejected in favor of more realistic and scientifically better based notions. The Upper Paleolithic cave art of Europe *is not* an art form endemic to caves—its location, which has led to much misinterpretation, is merely a product of taphonomic processes (Bednarik 1986, 1994a). It *is not* a record of the "origins of art"—much earlier paleoart exists elsewhere, and mostly outside Europe (Bednarik 1992b, 2003b). In fact there is far more "Middle Paleolithic" rock art surviving in the world than "Upper Paleolithic," and most of it occurs in Australia (Bednarik 1995, 2010).

Most of the world's Pleistocene rock art, probably over 99% of it, does not consist of semi-naturalistic animal figures, as the academic textbooks would tend to convey (Bednarik 1993, 1994b). Even in Europe's caves, zoomorphs, of which there are only a few thousand known, are outnumbered several times by noniconic marks. From the rest of the world we have almost no two-dimensional iconic depictions dating from this period. For instance, iconic (figurative) sculpture is well represented in the Upper Paleolithic of Russia and Siberia, but graphic (two-dimensional) art is limited almost entirely to non-figurative compositions (there are just two exceptions; Bednarik 1994b). Particularly notable are the sometimes incredibly complex engravings found on ivory and bone plaques. These exograms clearly had symbolic meanings; they often resemble maps but might well be mnemonic devices, e.g., for telling stories. Their meaning is inaccessible to us. Similarly, all credibly Pleistocene rock art of Australia, a massive corpus (Bednarik 2010), seems entirely noniconic, but can be shown to refer to very complex, if unexplained cultural practices. There is thus more surviving Mode 3 paleoart in the world than Mode 4, and with only one possible exception (Bednarik 2006), all of the known Middle (and Lower) Paleolithic graphic "art" appears to be noniconic, and of a quite narrow range of motif elements. (I must mention that the claim by Wendt [1974] of MSA iconic art from Namibia is no longer accepted; the painted plaques from Apollo 11 Cave are of the middle LSA.) Even the very limited Pleistocene paleoart of North America is entirely nonfigurative (Collins 2002; Collins et al. 1991, 1992; Robertson 1999). A possible exception are two apparent mammoth depictions from Utah (Malotki and Wallace 2011), but the time of extinction of proboscidians in North America is unclear.

In the Pleistocene, noniconic symbols (e.g., beads, geometric marks, hand signs, mnemonics) would have been the most economic form of storing cognitive information externally. The exploration of mark production to exploit their iconic ambiguity (Bednarik 2003c: 408, 412), by comparison, is comparatively trivial and may even have been regarded as a merely ludic pursuit. More importantly, its potential of holding cognitive information is relatively limited, and certainly less economic. If noniconic graphic art is cognitively more complex, why is iconic art largely absent from the early record? Perhaps people were able to produce iconographic markings, but rarely made use of this ability. To illuminate these issues I refer to my study of one of the very few remaining hunter-fisher-forager cultures of the world whose graphic art is entirely nonfigurative (Sreenathan et al. 2008; Bednarik and Sreenathan in press). The artistic production of the Jarawas of the Andaman

Islands, the traditional society that has most recently become available for detailed study (only since the late 1990s), reveals a purely nonfigurative tradition of graphic art. It consists of zigzags (*aawaav*), crosshatching (*aaweed*), parallel lines (*ikkaath* or *hechaya*), combined parallel lines and zigzags (*onebialile*), lozenge patterns (also called *onebialile*), crosses (*bethu oppo*), loops and circles (*bethubethu oppo*), fishbone pattern (*oppo haaneev*), and guilloches (*oppo diveel*). This limited motif repertoire has notable similarities with the currently available dominant repertoire of Pleistocene and in some cases very early Holocene graphic art, especially that of Asia (Bednarik 1993, 1994b). These include the reticulate arrangement of elements, the "enclosed spaces" aspect, the range of elemental forms (found also widely in Europe, Africa, and Australia), and the unique use of the rather distinctive guilloche (but known from the final Pleistocene of China; Bednarik 1992c).

However, Jarawas are perfectly capable of producing fine figurative drawings, especially when young (Sreenathan et al. 2008). Indeed, some of the graphic imagery of their young demonstrates outstanding observation of detail and talent (Fig. 6.4). If figurative drawing were perceived as ludic by early societies, as appears to be the case among the Jarawas, it may explain why there is no credible evidence that the graphic figurative art of the western European Upper Paleolithic is not the work of juveniles. It may also help explain the neoteny of anatomically modern humans.

One of the fundamental impairments in the epistemology of Pleistocene paleoart studies is that this field originates in Europe, and that it has been perceived as an

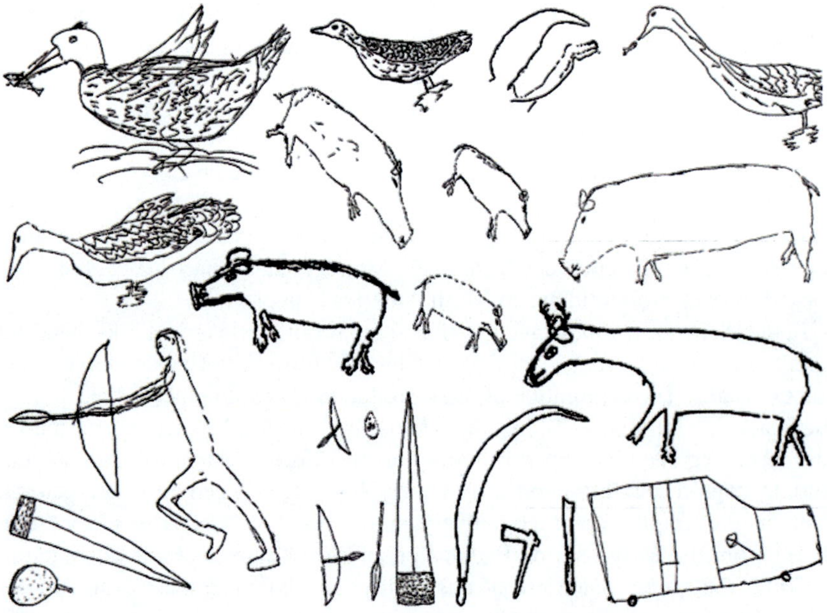

Fig. 6.4 Examples of the spontaneous figurative drawing skills of the Jarawa boy Enmay

essentially European issue for well over a century. While the discipline must be grateful for the leadership and pioneering excellence Europe has in many ways provided in this field, this bias has also resulted in various encumbrances. Had this field been pioneered by Islamic scholars, for instance—conditioned by the traditional avoidance of iconography—it might have attracted a very different epistemology. Burdened by theoretical impediments, European archeology has facilitated the establishment of countless false paradigms. It has ignored that there is far more Pleistocene rock art *outside* of Europe than there is in Europe; and that most surviving Pleistocene paleoart of the world is of *Middle* rather than *Upper* Paleolithic traditions (Bednarik 1986, 1992b, 2003b, 2010). Similarly, the figurative component of Franco-Cantabrian paleoart of the Upper Paleolithic has been overemphasized, which has led to several rather consequential sophisms. For instance, it was often assumed that "naturalistic" zoomorphs are the main themes of Pleistocene graphic traditions, which has led to sustained searches for such imageries across much of Eurasia (and even in North America in some cases) in efforts to locate Pleistocene paleoart. It has also facilitated the belief that these figurative elements of the Franco-Cantabrian rock and portable art were the more sophisticated elements in these traditions—yet another sophism.

Embedded in a contemporary Western mindset is a predisposition to perceiving cultural evolution as Darwinian, progressing from the "primitive" to the "developed." This clashes significantly with the observation that extant cultures in several continents would, under the European system, be defined as Paleolithic, Mesolithic, Neolithic, Bronze, and Iron Age, respectively. In some countries, all of these "stages" coexist today with modern society, which severely challenges the efficacy of this technological nomenclature in assessing, for instance, the cultural or cognitive complexity of such groups. The origins of paleoart have been traced mainly through archeological means, which have generally focused on "Paleolithic" cultures of one small region. These archeological finds consist of a vast number of art-like manifestations, including nonfigurative engravings, proto-figurines, pendants and beads, cupules, and linear petroglyphs. But when these are identified as being of Lower or Middle Paleolithic provenience (Bednarik 1992a, 2003b), archeology tends to reject them in compliance with its Darwinian dogma. The failure of this program is illustrated by many examples. For instance, Tasmanian culture, as observed ethnographically, is clearly of Mode 3 technological production (Foley and Lahr 1997), yet Tasmanians certainly created art-like exograms. Indeed, their rock art even resembles that of other "Middle Paleolithic" traditions (Bednarik et al. 2007), including in Europe (consider the sepulchral block in La Ferrassie; Peyrony 1934).

Another folklore attached to the European art corpus, that it marks the arrival of the "mythical moderns from Africa," has already been refuted in Chapter 2. On the basis of the current evidence the so-called Aurignacian, the technological tradition to which the invention of art had widely been attributed, is in all probability a tradition of robust humans, resembling so-called Neanderthals. My hypothesis replacing the replacement hypothesis, the domestication hypothesis, introduces a new, most intriguing possibility. Based on the observation that rapid gracilization is a universal feature of the last part of the Late Pleistocene, and that it marks effectively a fetalization of hominins, this hypothesis attributes the neoteny of "anatomically modern humans" to culturally moderated breeding patterns. The resulting loss of robusticity involved several reductions in evolutionary fitness. These deleterious effects occurred more or less concurrently in all regions occupied by humans at the time, including Australia, and cannot be explained in Darwinian terms.

Neoteny is in some respects certainly deleterious to a species, but it also involves evolutionary benefits. Most importantly, it facilitates the retention of plasticity or "morphological evolvability" (De Beer 1930: 93). As we have seen in the previous chapter, adaptively useful novelties become available as maturation genes are freed by pedomorphosis. In a species whose behavior is increasingly determined by cultural factors, corresponding plasticity of cultural behavior might foster the curiosity, inventiveness, and inquisitiveness of youth. It seems entirely possible that these traits, so important to the most recent cognitive developments of humans, may account for the rise of iconographic (i.e., "juvenile") graphic art forms, first pioneered by young people. A developing preference for iconic art toward the end of the Pleistocene would have had significant effects on the proliferation of new symbol systems; it made possible the revision of immutable constructs of reality expressed in the more regimented graphic semiotics of earlier societies. Hence it is possible to perceive the change to the more permutable figurative system as reflecting the trend toward neotenous attributes that marks the final Pleistocene. It is iconic representation that ultimately led to those systems of exograms we call writing. Once again, this interpretation of the record seems to provide a much fuller explanation than Merlin Donald's model. But just like it, it is also in need of considerable improvement and elaboration.

That is the thing about science: its Sisyphean task will never be completed.

References

Acharyya, S. K. and P. K. Basu 1993. Toba ash on the Indian subcontinent and its implications for the correlation of Late Pleistocene alluvium. *Quaternary Research* 40: 10–19.

Adams, F. and K. Aizawa 2001. The bounds of cognition. *Philosophical Psychology* 14: 43–64.

Adams, F. and K. Aizawa 2008. *The bounds of cognition*. Blackwell Publishing, Malden.

Aitchison, J. 1996. *The seeds of speech: language origin and evolution*. Cambridge University Press, Cambridge.

Aizawa, K. and F. Adams 2005. Defending non-derived content. *Philosophical Psychology* 18: 661–669.

Alemseged, Z., F. Spoor, W. H. Kimbel, R. Bobe, D. Geraads, D. Reed and J. G. Wynn 2006. A juvenile early hominin skeleton from Dikika, Ethiopia. *Nature* 443: 296–301.

Arensburg, B., L. A. Schepartz, A.-M.Tillier, B. Vandermeersch and Y. Rak 1990. A reappraisal of the anatomical basis for speech in the Middle Pleistocene hominids. *American Journal of Physical Anthropology* 83: 137–146.

Arensburg, B., A.-M.Tillier, B. Vandermeersch, H. Duday, L. A. Schepartz and Y. Rak 1989. A Middle Palaeolithic human hyoid bone. *Nature* 338: 758–760.

Armand, J. 1983. *Archaeological excavations in the Durkadi Nala—an early Palaeolithic pebble-tool workshop in central India*. Munshiram Manoharlal Publishers, Delhi.

Badam, G. L. and S. N. Rajaguru 1994. Comment on 'Toba ash on the Indian subcontinent and its implications for the correlation of Late Pleistocene alluvium' by S. K. Acharyya and P. K. Basu. *Quaternary Research* 41: 398–399.

Baddeley, A. 1992. Working memory. *Science* 255: 556–559.

Baddeley, A. 2000. The episodic buffer: a new component of working memory? *Trends in Cognitive Sciences* 4: 417–422.

Baron-Cohen, S. 1991. Precursors to a theory of mind: understanding attention in others. In A. Whiten (ed), *Natural theories of mind: evolution, development and simulation of everyday mindreading*, pp. 233–251. Basil Blackwell, Oxford.

Beaumont, P. B. in press. The edge: more on fire-making by 1.7 million years ago at Wonderwerk Cave in South Africa.

Bednarik, R. G. 1986. Parietal finger markings in Europe and Australia. *Rock Art Research* 3: 30–61, 159–170.

Bednarik, R. G. 1987. Engramme und Phosphene. *Zeitschrift für Ethnologie* 112(2): 223–235.

Bednarik, R. G. 1990a. On the cognitive development of hominids. *Man and Environment* 15(2): 1–7.

Bednarik, R. G. 1990b. *Homo erectus* skapade konst i mellanistiden. *Populär Arkeologi* 9(1): 18.

Bednarik, R. G. 1992a. On Lower Paleolithic cognitive development. In S. Goldsmith, S. Garvie, D. Selin and J. Smith (eds), *Ancient images, ancient thought: the archaeology of ideology*, pp. 427–435. Proceedings of the 23rd Annual Chacmool Conference, University of Calgary.

Bednarik, R. G. 1992b. Palaeoart and archaeological myths. *Cambridge Archaeological Journal* 2(1): 27–43.

Bednarik, R. G. 1992c. Palaeolithic art found in China. *Nature* 356, 116.

Bednarik, R. G. 1993. European Palaeolithic art—typical or exceptional? *Oxford Journal of Archaeology* 12(1): 1–8.

Bednarik, R. G. 1994a. A taphonomy of palaeoart. *Antiquity* 68: 68–74.

Bednarik, R. G. 1994b. The Pleistocene art of Asia. *Journal of World Prehistory* 8(4): 351–375.

Bednarik, R. G. 1995. Concept-mediated marking in the Lower Palaeolithic. *Current Anthropology* 36: 605–634.

Bednarik, R. G. 1997. The origins of navigation and language. *The Artefact* 20: 16–55.

Bednarik, R. G. 2002. Paläolithische Felskunst in Deutschland? *Archäologische Informationen* 25(1–2): 107–117.

Bednarik, R. G. 2003a. Seafaring in the Pleistocene. *Cambridge Archaeological Journal* 13(1): 41–66.

Bednarik, R. G. 2003b. The earliest evidence of palaeoart. *Rock Art Research* 20: 89–135.

Bednarik, R. G. 2003c. A figurine from the African Acheulian. *Current Anthropology* 44: 405–413.

Bednarik, R. G. 2006. The Middle Palaeolithic engravings from Oldisleben, Germany. *Anthropologie* 44(2): 113–121.

Bednarik, R. G. 2008. Children as Pleistocene artists. *Rock Art Research* 25: 173–182.

Bednarik, R. G. 2010. Pleistocene rock art in Australia. *Anthropos* 105(1): 3–12.

Bednarik, R. G., G. Andrews, S. Cameron and E. Bednarik 2007. Petroglyphs of Meenamatta, the Blue Tier mountains, Tasmania. *Rock Art Research* 24: 161–170.

Bednarik, R. G., G. Kumar, A. Watchman and R. G. Roberts 2005. Preliminary results of the EIP project. *Rock Art Research* 22: 147–197.

Bednarik, R. G. and M. Sreenathan in press. Traces of the ancients: ethnographic vestiges of Pleistocene art.

Berger, L. R., S. E. Churchill, B. De Klerk and R. L. Quinn 2008. Small-bodied humans from Palau, Micronesia. *PLoS ONE* 3(3): e1780. DOI: 10.1371/journal.pone.0001780.

Bickerton, D. 1990. *Language and species*. University of Chicago Press, Chicago, IL.

Bickerton, D. 1993. Putting cognitive carts before linguistic horses. *Behavioural and Brain Sciences* 16: 749–750.

Bickerton, D. 1996. *Language and human behaviour*. UCL Press, London.

Bickerton, D. 2010. *Adam's tongue: how humans made language, how language made humans*. Hill and Wang, New York, NY.

Block, N. 2005. Review of Alva Noë. *Journal of Philosophy* 102: 259–272.
Brace, C. L. 1993. 'Popscience' versus understanding the emergence of the modern mind. Review of *Origins of the Modern Mind: Three Stages in the Evolution of Culture and Cognition*, by Merlin Donald, in *Behavioral and Brain Sciences* 16(4): 750–751.
Brace, C. L. 1996. Racialism and racist agendas: review of *Race, evolution and behavior: a life history perspective*, by J. Philippe Rushton. *American Anthropologist* 98(1): 176–177.
Brace, C. L. 1999. An anthropological perspective on 'race' and intelligence: the non-clinal nature of human cognitive capabilities. *Journal of Anthropological Research* 55(2): 245–264.
Bryant, E. H., S. A. McComas and L. M. Combs 1986. The effect of an experimental bottleneck on quantitative genetic variation in the housefly. *Genetics* 114: 1191–1211.
Chauhan, P. R. and R. Patnaik 2008. The Narmada Basin Palaeoanthropology Project in central India. *Antiquity* 82(317): http://antiquity.ac.uk/projgall/chauhan/index.html
Ciochon, R. 1995. The earliest Asians yet. *Natural History* 104(12): 50–54.
Clottes, J. 1985. Conservation des traces et des empreintes. *Histoire et Archéologie* 90: 40–49.
Clottes, J. 1986. Comment on R. G. Bednarik, 'Parietal finger markings in Europe and Australia'. *Rock Art Research* 3: 160–161.
Clottes, J. 1997. *Niaux. Die altsteinzeitlichen Bilderhöhlen in der Ariège* (transl. French edn 1995). Jan Thorbecke Verlag, Sigmaringen.
Clottes, J. and J. Courtin 1995. *La grotte Cosquer: peintures et gravures de la caverne engloutie*. Seuil, Paris.
Collins, M. B. 2002. The Gault Site, Texas, and Clovis research. *Athena Review* 3(2): 31–42, 100–101.
Collins, M. B., T. R. Hester and P. J. Headrick 1992. Engraved cobbles from the Gault Site, central Texas. *Current Research in the Pleistocene* 9: 3–4.
Collins, M. B., T. R. Hester, D. Olmstead and P. J. Headrick 1991. Engraved cobbles from early archaeological contexts in central Texas. *Current Research in the Pleistocene* 8: 13–15.
Culotta, E. 1995. Asian anthropoids strike back. *Science* 270: 1116–1117.
Cummins, R. 1996. *Representations, targets, and attitudes*. MIT Press, Cambridge.
Cynx, J. and S. J. Clark 1993. Ethological foxes and cognitive hedgehogs. *Behavioural and Brain Sciences* 16: 756–757.
Dawkins, R. 1976. *The selfish gene*. Oxford University Press, Oxford.
Dean, D. and E. Delson 1995. *Homo* at the gates of Europe. *Nature* 373: 472–473.
De Beer, G. R. 1930. *Embryology and evolution*. Oxford University Press, Oxford.
de Lumley, M.-A. and A. Sonakia 1985. Première découverte d'un *Homo erectus* sur le continent indien, à Hathnora, dans le moyenne vallée de la Narmada. *L'Anthropologie* 89: 13–61.
Donald, M. 1991. *Origins of the modern mind: three stages in the evolution of culture and cognition*. Harvard University Press, Cambridge, MA.
Donald, M. 1993. On the evolution of representational capacities. *Behavioural and Brain Sciences* 16: 775–785.
Donald, M. 2001. *A mind so rare: the evolution of human consciousness*. W. W. Norton, New York, NY.
Dretske, F. 1981. *Knowledge and the flow of information*. MIT Press, Cambridge.
Dubois, E. 1894. *Pithecanthropus erectus, eine menschenähnliche Übergangsform aus Java*. Landersdrucherei, Batavia.
Duday, H. and M.-A. Garcia 1983. Les empreintes de l'homme préhistorique: la grotte du Pech-Merle à Caberets (Lot). *Bulletin de la Société Française* 8: 205–215.
Duday, H. and M.-A. Garcia 1985. L'homme et la caverne. *Histoire et Archéologie* 90: 35–39.
Duday, H. and M.-A. Garcia 1990. L'ichnologie ou la mémoire des roches. In N. Navialoff, R. Jaffard and P. Brenot (eds), *Le concept de mémoire*, pp. 55–66. Hartman, Paris.
Dunbar, R. 1996. *Grooming, gossip and the evolution of language*. Faber and Faber, London.
Enard, W., M. Przeworski, S. E. Fisher, C. S. Lai, V. Wiebe, T. Kitano, A. P. Monaco and S. Pääbo 2002. Molecular evolution of FOXP2, a gene involved in speech and language. *Nature* 418: 869–872.

Everett, D. 2005. Cultural constraints on grammar and cognition in Pirahã: another look at the design features of human language. *Current Anthropology* 46: 621–646.

Fairservice, W. A., Jr. 1975. *The threshold of civilization*. Scribner, New York, NY.

Falk, D. 1975. Comparative anatomy of the larynx in man and chimpanzee: implications for language in Neanderthal. *American Journal of Physical Anthropology* 43: 123–132.

Falk, D. 1987. Hominid paleoneurology. *Annual Review of Anthropology* 16: 13–30.

Falk, D. 2009. *Finding our tongues: mothers, infants and the origins of language*. Basic Books, New York, NY.

Fodor, J. 1990. *A theory of content and other essays*. MIT Press, Cambridge.

Foley, R. and M. M. Lahr 1997. Mode 3 technologies and the evolution of modern humans. *Cambridge Archaeological Journal* 7: 3–36.

Gabunia, L. and A. Vekua 1995. A Plio-Pleistocene hominid from Dmanisi, East Georgia, Caucasus. *Nature* 373: 509–512.

Garcia, M.-A. 2003. Prints of traces of humans and animals. In J. Clottes (ed), *Chauvet Cave: the art of earliest times*, pp. 34–43. University of Utah Press, Salt Lake City.

Garcia, M.-A. and H. Duday 1993. Les empreintes de mains dans l'argile des grottes ornées. *La main dans la préhistoire. Dossiers d'Archéologie* 178: 56–59.

Gregory, R. L. 1970. *The intelligent eye*. Weidenfeld and Nicolson, London.

Guthrie, R. 2005. *The nature of Paleolithic art*. The University of Chicago Press, Chicago, IL/London.

Guzder, S. 1980. *Quaternary environments and Stone Age cultures of the Konkan, coastal Maharashtra, India*. Deccan College, Pune.

Huang W., R. Ciochon, G. Yumin, R. Larick, F. Qiren, H. Schwarcz, C. Yonge, J. De Vos and W. Rink 1995. Early *Homo* and associated artefacts from Asia. *Nature* 378: 275–278.

Huang W. and F. Qiren 1991. *Wushan hominid site*. Ocean Press, Beijing.

Hurcombe, L. 2004. The stone artefacts from the Pabbi Hills. In R. W. Dennell (ed), *Early hominin landscapes in northern Pakistan: investigations in the Pabbi Hills*, pp. 222–292. International Series 1265, British Archaeological Reports, Oxford.

Jacques, S. and P. D. Zelazo 2005. Language and the development of cognitive flexibility: implications for theory of mind. In J. W. Astington and J. A. Baird (eds), *Why language matters for theory of mind*, pp. 144–162. Oxford University Press, Toronto.

Khatri, A. P. 1963. Mahadevian: an Oldowan pebble culture of India. *Asian Perspectives* 6: 186–197.

Klaatsch, H. 1908. The skull of the Australian Aboriginal. *Report from the Pathological Laboratory of the Lunacy Department, New South Wales* 1: 43–167.

Krause, J., C. Lalueza-Fox, L. Orlando, W. Enard, R. E. Green, H. A. Burbano, J.-J. Hublin, C. Hänni, J. Fortea, M. de la Rasilla, J. Bertranpetit, A. Rosas and S. Pääbo 2007. The derived FOXP2 variant of modern humans was shared with Neandertals. *Current Biology* 17(21): 1908–1912.

Laitman, J., J. S. Reidenberg and P. J. Gannon 1992. Fossil skulls and hominid vocal tracts: new approaches to charting the evolution of human speech. In J. Wind, B. Chiarelli, B. Bichakjian and A. Nocentini (eds), *Language origin: a multidisciplinary approach*, pp. 395–407. Kluwer, Dordrecht

Larnach, S. L. and N. W. G. Macintosh 1974. A comparative study of Solo and Australian Aboriginal crania. In A. P. Elkin and N. W. G. Macintosh (eds), *Grafton Elliot Smith: the man and his world*, pp. 95–102. Sydney University Press, Sydney.

Lashley, K. 1950. In search of the engram. *Society of Experimental Biology* Symposium 4: 454–482.

Lieberman, P. and E. S. Crelin 1971. On the speech of Neanderthal man. *Linguistic Inquiry* 2: 203–222.

Lieberman, P., E. S. Crelin and D. H. Klatt 1972. Phonetic ability and related anatomy of the newborn, adult human, Neanderthal man, and the chimpanzee. *American Anthropologist* 74: 287–307.

Lieberman, P., J. R. Laitman, L. S. Reidenburg and P. J. Gannon 1992. The anatomy, physiology, acoustics and perception of speech: essential elements in analysis of the evolution of human speech. *Journal of Human Evolution* 23: 447–467.

Malafouris, L. 2004. The cognitive basis of material engagement: where brain, body and culture conflate. In E. DeMarrais, C. Gosden and C. Renfrew (eds), *Rethinking materiality: the engagement of mind with the material world*, pp. 53–62. McDonald Institute for Archaeological Research, Cambridge.

Menary, R. 2007. *Cognitive integration: mind and cognition unbounded*. Palgrave Macmillan, Hampshire.

Miller, G. A. 1956. The magical number seven, plus or minus two: some limits on our capacity for processing information. *Psychological Review* 63: 81–97.

Mishra, S. 1992. The age of the Acheulian in India: new evidence. *Current Anthropology* 33: 325–328.

Misra, S. and S. N. Rajaguru 1994. Comment on 'Toba ash on the Indian subcontinent and its implications for the correlation of Late Pleistocene alluvium' by S. K. Acharyya and P. K. Basu. *Quaternary Research* 41: 396–397.

Paddayya, K., B. A. B. Blackwell, R. Jhaldiyal, M. D. Petraglia, S. Fevrier, D. A. Chaderton II, J. I. B. Blickstein and A. R. Skinner 2002. Recent findings on the Acheulian of the Hunsgi and Baichbal valleys, Karnataka, with special reference to the Isampur excavation and its dating. *Current Science* 83(5): 641–647.

Pales, L. 1954. Les empreintes de pieds humains de la Tana della Basura (Toirano). *Revue d'Etudes Ligures* 1: 1–12.

Pales, L. 1960. Les empreintes de pieds humains de la 'Grotta della Basura'. *Revue d'Etudes Ligures* 4: 25–90.

Pales, L. 1976. Les empreintes de pieds humains dans ls cavernes. *Archives de l'Institut de Paléontologie Humaine, Paris* 36: 1–166.

Peyrony, D. 1934. La Ferrassie. Moustérien, Périgordien, Aurignacien. *Préhistoire* 3: 1–92.

Premack, D. 2007. Human and animal cognition: continuity and discontinuity. *Proceedings of the National Academy of Sciences of the United States of America* 104: 13861–13867.

Prinz, J. 2006. Putting the brakes on enactive perception. *Psyche* 12: 1–19.

Rendell, H. M., R. W. Dennell and M. A. Halim 1989. *Pleistocene and Palaeolithic investigations in the Soan Valley, northern Pakistan. British Archaeological Mission to Pakistan, Series 2*. International Series 544, British Archaeological Reports, Oxford.

Robertson, T. 1999. Symbolic and notational expressions of central Texas Palaeo-Indian culture: engraved artifacts of the Gault Site. *McNair Research Journal* 3: 123–136.

Roveland, B. 2000. Footprints in the clay: Upper Paleolithic children in ritual and secular contexts. In J. Sofaer Derevenski (ed), *Children and material culture*, pp. 29–38. Routledge, London.

Rowland, M. 2003. Comment on R. G. Bednarik, 'Seafaring in the Pleoistocene'. *Cambridge Archaeological Journal* 13(1): 52–54.

Rupert, R. 2004. Challenges to the hypothesis of extended cognition. *The Journal of Philosophy* 101: 1–40.

Sanjuan, J., A. Tolosa, J. C. Gonzalez, E. J. Aguilar, J. Perez-Tur, C. Najera, M. D. Molto and R. Frutos 2006. Association between FOXP2 polymorphisms and schizophrenia with auditory hallucinations. *Psychiatric Genetics* 16: 67–72.

Schwartz, J. H. and I. Tattersall 1996. Whose teeth? *Nature* 381: 201–202.

Semon, R. 1904. *Die Mneme*. W. Engelmann, Leipzig.

Semon, R. 1921. *The mneme*. George Allen & Unwin, London.

Sharma, H. C. and S. K. Roy 1985. On the discovery of a pebble-tool industry in the Garo Hills, Meghalaya. In V. N. Misra and P. Bellwood (eds), *Recent advances in Indo-Pacific prehistory*, pp. 89–91. Oxford and IBH Publishing, New Delhi.

Sharpe, K. and L. Van Gelder 2006. Finger flutings in Chamber A1 of Rouffignac Cave, France. *Rock Art Research* 23: 179–198.

Sreenathan, M., V. R. Rao and R. G. Bednarik 2008. Palaeolithic cognitive inheritance in aesthetic behavior of the Jarawas of the Andaman Islands. *Anthropos* 103: 367–392.

Swisher, C. C., G. H. Curtis, T. Jacob, A. G. Getty, A. Suprijo and Widiasmoro 1994. The age of the earliest hominids in Indonesia. *Science* 263: 1118–1121.

Tarnow, E. 2009. Short term memory may be the depletion of the readily releasable pool of presynaptic neurotransmitter vesicles of a metastable long term memory trace pattern. *Cognitive Neurodynamics* 3(3): 263–269.

Thorne, A. G. 1980. The longest link: human evolution in Southeast Asia and the settlement of Australia. In J. J. Fox, R. G. Garnaut, P. T. McCawley and J. A. C. Mackie (eds), *Indonesia: Australian perspectives*, pp. 35–43. Research School of Pacific Studies, Australian National University, Canberra.

Thorne, A. G. 1989. *Man on the rim: the peopling of the Pacific*. Angus and Robertson, Sydney.

Thorne, A. G. and S. R. Wilson 1977. Pleistocene and recent Australians: a multivariate comparison. *Journal of Human Evolution* 6: 394–402.

Thorne, A. G. and M. H. Wolpoff 1981. Regional continuity in Australasian Pleistocene hominid evolution. *American Journal of Physical Anthropology* 55: 337–349.

Tindale, N. 1962. Some population changes among the Kaiadilt of Bentinck Island, Queensland. *Records of the South Australian Museum* 14(2): 297–336.

Weidenreich, F. 1943. The skull of *Sinanthropus pekinensis*: a comparative study of a primitive hominid skull. *Palaeontologica Sinica* D10: 1–485.

Weidenreich, F. 1945. The Keilor skull: a Wadjak type from south-east Australia. *American Journal of Physical Anthropology* 3: 225–236.

Weidenreich, F. 1951. Morphology of Solo man. *Anthropological Papers of the American Museum of Natural History* 43(3): 205.

Wendt, W. E. 1974. Art mobilier aus der Apollo 11 Grotte in Südwest-Afrika. *Acta Praehistorica et Archaeologica* 5: 1–42.

Wolpoff, M. H. 1980. *Paleoanthropology*. Alfred A. Knopf, New York, NY.

Wolpoff, M. H. 1989. Multiregional evolution: the fossil alternative to Eden. In P. Mellars and C. Stringer (eds), *The human revolution: behavioural and biological perspectives on the origins of modern humans*, pp. 62–108. Edinburgh University Press, Edinburgh.

Wolpoff, M. H. 1991. Theories of modern human origins. In G. Bräuer and F. H. Smith (eds), *Continuity or replacement? Controversies in* Homo sapiens *evolution*, pp. 25–64. A. A. Balkema, Rotterdam.

Wolpoff, M. H. 1997. *Human evolution*. McGraw-Hill, New York, NY.

Wood, B. A. and A. Turner 1995. Out of Africa and into Asia. *Nature* 378: 239–240.

Zhang, J., D. M. Webb and O. Podlaha 2002. Accelerated protein evolution and origins of human-specific features: FOXP2 as an example. *Genetics* 162: 1825–1835.

Chapter 7
Advanced Human Cognition: A Faustian Deal

A Little Neuroscience

As stated earlier in this book, in seeking to determine how we became human, our search would benefit from an initial focus on those rather few characteristics that appear to be uniquely and quintessentially human. Preoccupied with inconclusive word games, the social sciences have not produced much sound empirical data. Neuroscience, on the other hand, has, but that wealth remains largely untapped by those examining hominin cognitive evolution. For instance, one of the major differences between humans and other extant primates is found in Brodmann's area 10 (Brodmann 1912) in the prefrontal lobe (Fig. 7.1), apparently much more developed in humans than in chimpanzees (Semendeferi 2001; Diller et al. 2002). This cortical region supports higher cognitive functions, including the extraction of meaning from experience; the organization of mental contents that control creative thinking and language; artistic creation; initiation of expression of and planning for future action (Damasio 1985). Hodgson and Helvenston (2006) suggest that area 10 is one of the most likely substrates for the expansion of complex, sustained, and focused human consciousness, which is one of the major differences between humans and other extant primates. Details of their neurological disparities suggest that distinctive changes must have taken place during the course of hominin evolution, since the phylogenetic split in the Miocene period.

Differences between the brains of humans and members of the pongid clade are essentially in respect of *structure* and *size*. We consider the structural dissimilarities first.

Specific neurons, called VENs (von Economo neurons), occur in both apes (and other mammals) and humans (Nimchimsky et al. 1999; Watson et al. 2006), but they are larger and far more numerous in the latter, occurring in humans in the anterior cingulate cortex (Allman et al. 2002; Hayashi 2006) and the frontoinsular cortex (Sridharan et al. 2008). These structures are thought to be involved in complex social emotion and cognition. Nerve cells in humans, for example, in the primary visual cortex, are arranged in far more complex patterns than in apes (Preuss and Coleman 2002). The minicolumn, the mammalian brain's basic information processing structure, in the left planum temporale, is significantly enlarged in the human, relative to the chimpanzee or rhesus monkey (Buxhoeveden and Casanova

R.G. Bednarik, *The Human Condition*, Developments in Primatology: Progress and Prospects, DOI 10.1007/978-1-4419-9353-3_7, © Springer Science+Business Media, LLC 2011

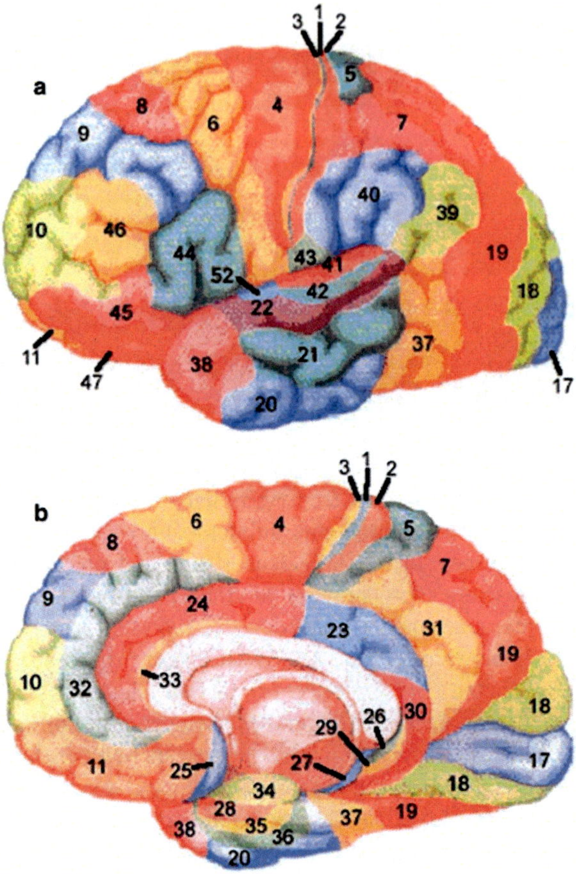

Fig. 7.1 (**a**) The lateral surface of the brain's left hemisphere showing the numbers signifying Brodmann's areas. (**b**) The mesial surface of the right hemisphere with Brodmann's areas indicated by number

2002), containing the axons, dendrites, and synapses that make neural connections (Sherwood et al. 2009). Significantly, the planum temporale is involved in language production. Ullian et al. (2001) found that synapses form between neurons only in the presence of astrocytes, neuroglial cells constituting almost half the cells in a human brain (see also Ullian et al. 2004; Barres 2008). These cells secrete the protein thrombospondin, which triggers synapse formation (Christopherson et al. 2005) and of which the human brain produces about six times as much as that of chimpanzees or macaques (Cáceres et al. 2007). It also produces about twice as much of THBS4 and THBS2 messenger RNA (mRNA), respectively, in the human cerebral cortex. (RNAs are ribonucleic acids that, like DNA, can carry genetic information.) Thrombospondin expression differences were observed in the forebrain (cortex and caudate), whereas the cerebellum and most nonbrain tissues exhibit similar levels of the two mRNAs, in humans and chimpanzees. Increased expression of thrombospondins in human brain evolution could result in changes in synaptic organization and plasticity, and contribute to the distinctive cognitive abilities of humans, as well as to the vulnerability to neurodegenerative disease that seems unique to humans (Walker and Cork 1999; Olson and Varki 2003).

VENs are considered to participate in rapid signal transmissions and are relatively newly evolved in mammals, being present in humans, pongids, sperm and beluga whales, bottlenosed and Risso's dolphins, and African and Asian elephants (Coghlan 2006; Hof and Van der Gucht 2007; Butti et al. 2009; Hakeem et al. 2009; Seeley et al. 2006). This list suggests von Economo cells may be restricted to relatively large animals with large brains and extensive social networks.

Those areas of the anterior cingulate cortex that contain VENs have been proposed to be a phylogenetically new specialization of the neocortex (Allman et al. 2001) rather than a more primitive state of cortical evolution as most other areas of the cingulate cortex (Nimchinsky et al. 1995). The anterior cingulate is an area involved with a variety of emotions, both positive and negative, and its VENs are believed to project to Brodmann's area 10 (Allman et al. 2002). While the human frontal lobes are not relatively oversized, parts such as areas 10 and 13 may be enlarged relative to what would be expected from a primate of human body size. The increase in area 10, used in retrieving memories from the individual's past experience and in planning future action, has been suggested to begin with *Homo habilis* (Semendeferi 1994). Area 13 is part of the limbic system (Heimer et al. 2008) and involved in emotional, motivational, and social behavior via its interconnections with other limbic and cortical structures. The cingulate gyrus, a significant part of the limbic lobe, is of substantial size in humans. It subsumes viscero-motor, cognitive-effector, instant emotional experiences, adaptive motor responses, and sensory processing areas (Mega and Cummings 1997). The extended and in some ways unique human limbic system (Heimer et al. 2008) includes the hippocampus and the amygdala, the septum, olfactory nucleus, entorhinal cortex, bed nucleus of the stria terminalis, and the nucleus basalis of Meynert. The amygdala is implicated in a spectrum of social attributions, such as appraisal of the emotional state of others (Adolphs et al. 1994), value judgments such as trustworthiness (Adolphs et al. 1999), and the emotional tone of memory consolidation and restructuring. The hippocampus receives its input from the entorhinal cortex, which receives its inputs from the associative neocortex—the most recent cortical development, which is involved in spatial orientation (Frank et al. 2000). The entorhinal cortex, hippocampus, and amygdala are involved in memory functioning (Gloor 1990, 1992). Since there are few conditions that involve almost exclusively the amygdala (e.g., Urbach-Weithe disease) its now accepted neuroanatomical extension includes the subpallidal region and the bed nucleus of the stria terminalis.

The prefrontal cortex as well as portions of the posterior association areas have in humans become enlarged beyond what would be expected in comparison to primary sensorimotor structures (Preuss 2001). Preuss and Kaas (1999) report that the human Brodmann's area 17, comprising the primary visual areas, differs from both apes and monkeys in the way information is segregated from the magnocellular and parvocellular layers of the lateral geniculate nucleus. These authors attribute to humans an improved ability of evaluating moving stimuli. Holloway (1995, 1996) notes that the visual striate cortex and lateral geniculate are significantly smaller in a human than expected for an ape of human size and suggests that the variation in the former area began with australopithecines (Holloway 2001). Another distinctive difference is in the pyramidal motor system, the most recently evolved part of the motor cortex. In particular, the rostral section of the motor cortex is phylogenetically recent. In contrast to apes, where the motor cortex is located on both banks of the central sulcus, it is in humans located in front of the central sulcus. The caudal primary motor area is mediated by corticospinal efferents in the extrapyramidal system, which includes the caudate and putamen, i.e., the striatum. The latter and the globus pallidus form the basal ganglia, thought to integrate our emotion and reason, generating motor neuron activity and motor output (Rathelot and Strick 2009).

Another difference between pongid and human brains concerns the cerebellum, which is smaller in humans than would be expected in an ape of human body size, but larger than expected in the gorilla (Semendeferi 2001). It serves fine motor tuning, balance, and some aspects of cognition (Leiner et al. 1995), in the routinization of complex cognitive procedures, error detection, and language. The ability to predict the actions of others and the preparation for behavioral responses have also been attributed to cerebellar learning of sequences (Mueller and Courchesne 1998).

The parietal association area is larger in humans than in apes at the expense of the occipital cortices. The planum temporale, presenting a left-right asymmetry favoring the left (Geschwind and Levitsky 1968), has been related to language reception, but it is also present in apes (Gannon et al. 1998, 2001; Diller et al. 2002). This challenges the often-perceived simple relationship between this asymmetry and language. These similarities are perhaps merely homologous in apes and humans, as suggested by the work of Stepniewska et al. (1993) with owl monkeys. It showed that when the homologue of Broca's area is stimulated in that species it produces oral and laryngeal responses. The insula in the anterior Sylvian fissure of the temporal lobe has also been suggested to be large in humans, constituting part of the extended limbic system (Heimer et al. 2008). The nucleus subputaminalis in the basal forebrain is unique in humans, providing cholinergic innervation to the inferior frontal gyrus where Broca's area (crucial for speech) is located. Wernicke's area is specialized in humans for the reception of sounds, especially language comprehension. Located in the posterior temporal lobe, this area has six layers, in contrast to the three-layered allocortex (hippocampus and olfactory cortex) (Buxhoeveden et al. 1996).

Having thus considered *structural* differences between human and ape brains, we now turn to the question of their respective *sizes*. Encephalization is one of the two most distinguishing characteristics of humans, the other being neotenization (as discussed in Chapters 5 and 6). At 1350 ml, the human brain is significantly larger than that of any other species, relative to body size. This is perhaps best expressed by the encephalization quotient (EQ), introduced by Jerison (1973). It expresses the ratio of *actual* brain volume to "*expected* brain volume", the latter being based on average sizes of living mammals (Kolb and Whishaw 2008: 41). It predicts an increase of 0.75 in brain size for every unit of body size increase (Martin 1996). Based on the cat having an EQ of 1.0, the quotient thus reflects the increase in brain size over and beyond that demanded by body size. The rhesus monkey's EQ of 2.09

(200 ml brain volume) is not spectacularly smaller than the EQ of the chimpanzee of 2.48, with more than double the brain volume (440 ml). The chimpanzee is our closest living relative, genetically much closer to us than the gorilla, but when it comes to the EQ, it is decidedly dwarfed by our EQ of 7.30.

Since the cell bodies of large brains are more scattered, there is more room in them for interconnections between areas (Semendeferi 2001). It is precisely the expansion of association cortices that has made the human brain disproportionately large (Preuss 2000). The human cortex is ten times larger than that of the macaque and three to four times larger than any ape's (Semendeferi 2001: 108). It had long been assumed that the frontal lobes in humans were much larger than would be expected for a primate of human body size (Brodmann 1912; Blinkov and Glezer 1968), but recent evidence suggests that they are just about what should be expected (Semendeferi 1997, 2001). However, as noted above, specific areas (e.g., expanded limbic system, Brodmann's areas 10 and 13) are larger than would be predicted.

In humans, adult brain size is partly determined by the number of neurons produced and retained during an individual's ontogeny. Different neural areas differ in the length of the embryonic period of neuronal cytogenesis of precursor cells. The longest periods of cytogenesis apply to the areas of greatest degree of enlargement (cortex), and the shortest apply to the least enlarged brain areas (medulla). This suggests that brain size as well as relative size of specific structures could be the result of developmental timing (Kaskan and Finley 2001). A decrease in rates of apoptosis (the "normal", programmed death of neurons or neuronal precursor cells) could also account for increased brain size. Rakic and Kornack (2001) have suggested that in encephalization, cortical neurons are formed from precursor cells lining the ventricles and migrating radially along glial fascicles to reach their cortical destination. Thus, the neurons that are formed first make up the deepest layers of the neocortex, and those formed last are its most superficial. Since the neocortical thickness of the macaque brain is similar to the human brain, it is significant that the surface area of the human cortex is ten times as large as that of the macaque. Surface area is determined by the number of these radial columns, while the number of cells in each column determines cortex thickness. It is suggested, from comparing macaques and mice, that the process is regulated by the numbers of mitotic cell divisions involved in the cytogenesis of neurons prior to or during the early stages of the initial formation of radial columns.

It is thus currently assumed that relatively minor changes in developmental processes, which might be controlled by only a few genes that alter cytogenesis, could account for variations in brain size among different species.

The Susceptibility of the Human Brain to Illness

This digression reviewing the differences between the brains of some primates leads us to the key observation that extant nonhuman primates appear to be largely free of the neurodegenerative diseases as well as numerous genetic defects that are so prominent among modern man (Rubinsztein et al. 1994; Walker and Cork 1999;

Olson and Varki 2003; Bailey 2006). A review of the relevant brain illnesses also suggests that they involve largely the very same areas of the brain that are the phylogenetically most recent, in that they differ most from those of other extant primates. It is these areas that support the much-vaunted perceived advanced cognitive and intellectual characteristics of modern humans. This raises a fascinating perspective of the "human condition", the subject of this book: it appears that *Homo sapiens sapiens* is paying a price for his extraordinary abilities (Crow 1997; Randall 1998).

Not only have the mental and cognitive developments in our brain rendered humans vulnerable to neurodegenerative diseases as well as frontal lobe connectivity problems—such as those causing autism, schizophrenia, bipolar illness, demyelination or dysmyelination—which significantly other primates seem to be free of; they have apparently also facilitated the rise of quite specific personality disorders, such as obsessive compulsive disorder and sociopathic or antisocial personality disorders. Like Rett and Down syndromes and literally thousands of other known genetic impairments endemic to humans, they could be expected to have been vigorously selected against by natural evolution. The same applies to such conditions as chronic fatigue syndrome, which, although of unknown etiology, is often related to psychiatric disorders (Brown et al. 2010). Here, however, I will focus on neurodegenerative pathologies, of which a considerable variety is being distinguished. Among the better known are Alzheimer's, Huntington's, and Parkinson's diseases, frontotemporal dementia, and behavioral variant FTD; but amyotrophic lateral and diffuse myelinoclastic sclerosis, AIDS dementia, Batten disease and neuronal ceroid lipofuscinosis, Creutzfeldt-Jakob disease, and many others also pertain. Here I will very briefly review the first listed, as well as several further common human brain illnesses posing evolutionary selection issues: obsessive compulsive disorder, bipolar or manic-depressive disorder, schizophrenia, multiple sclerosis, autism spectrum disorders, Asperger's syndrome, temporal lobe epilepsy, and middle cerebral artery stroke.

Alzheimer's disease (AD) can be diagnosed in people from the 30s onward, but is primarily a disease of aging. It results from extracellular plaque deposition of beta amyloid and intracellular accumulation of tau, a protein. Tau is a component of intracellular neurofibrillary tangles. These plaques and tangles, clearly visible with MRI, are initially found primarily in the hippocampus and entorhinal cortex, later in some areas of the frontal cortex and temporal (medial temporal lobe), and parietal association cortex. AD targets the limbic structures (Hyman et al. 1984), including the amygdala, the locus coeruleus, and the cholinergic neurons of the nucleus basalis of Meynert. As a result of the plaque deposition, neurons and synapses die, axons degenerate, and connections are lost; general atrophy of the cortex and brain shrinkage occur (Smith 2002). VENs are particularly vulnerable to AD, and about 60% of them may be lost in the anterior cingulate cortex (Seeley et al. 2006).

VENs are also highly implicated in *frontotemporal dementia* (FTD, also known as Pick's disease) (Mayo Clinic 2010). This group of relatively rare disorders affects primarily the frontal and temporal lobes, associated with personality, behavior, and language. In one variant of frontotemporal dementia known as behavioral variant bvFTD, both the anterior cingulate cortex and the orbital frontoinsula show marked signs of focal degeneration, which is prominent in the right hemisphere (Seeley et al. 2007).

Huntington's disease (HD) derives from cell loss in the basal ganglia and cortex (Revilla and Grutzendle 2008). This movement, cognitive, and behavioral disorder can affect most age groups and occurs in the neostriatum, where marked atrophy of the caudate and putamen is accompanied by selective neuronal loss and astrogliosis. Degrees of atrophy in other regions, including the globus pallidus, thalamus, subthalamic nucleus, substantia nigra, and cerebellum, depend upon the progress of the disease. Its genetic basis involves the expansion of a cysteine-adenosine-guanine (CAG) repeat encoding, a polyglutamine tract in the N-terminus of the protein product called Huntingtin—the function of which remains unknown.

Parkinson's disease (PD) is associated with a loss of dopaminergic nigrostriatal neurons, which are located in the substantia nigra of the midbrain. By the time a patient is diagnosed with PD, usually at stage three of six defined stages, about 60–70% of the substantia nigra dopamine cells are already lost. By that time the substantia nigra, basal ganglia, amygdala, part of the limbic system, nucleus basalis of Meynert, and part of the extended amygdala have all been affected. Recent research suggests that *Helicobacter pylori*, a bacterium implicated in ulcers and stomach cancer, may help trigger PD.

Obsessive compulsive disorder (OCD) is of particular interest here, as we will see later. The persistent thoughts, feelings, and impulses characterizing this syndrome are attributable to an overactive inferior prefrontal cortex. Appearing first in childhood to early adulthood, this anxiety disorder affects areas of the brain called the "worry circuit" and is connected with an imbalance of the neurotransmitter serotonin (Schwartz and Begley 2002). Excessive activity in the inferior prefrontal cortex leads to the development of obsessive stereotypical behaviors. The striatum (caudate nucleus and putamen) is also overactivated in OCD sufferers. Projecting to the striatum, the inferior prefrontal cortex, orbitofrontal cortex, and the cingulate cortex cause the caudate to be overactive in the striosome area, thus bringing emotional tones and valences into the experience via the amygdala because it also projects into this same striosome area. Between the matrisome and the striosome areas are the tonically active neurons (TANs), which integrate the input from the inferior orbital frontal cortex via the striosomes with the input from the amygdala and orbitalfrontal region, also via the striosomes. The TANs thus function as a gating mechanism between the matrisome and the striosome regions.

The direct route by which the striatum projects to the cortex is via the globus pallidus, thalamus, and motor and premotor cortex. The indirect route neurons project from the striatum to the globus pallidus, the subthalamic nucleus to the thalamus, and then to motor and premotor cortex. The direct pathway from the striatum calms the cortex, whereas the indirect stimulates it, and the gating TANs determine which pathway is taken. When working effectively, the TANs modulate the orbital frontal cortex and anterior cingulate by adjusting the degree to which the thalamus drives both areas. But when the system is faulty, as in OCD, the error detector centered in the orbital frontal cortex and anterior cingulate can be locked into a pattern of repetitive firing. This triggers in the patient an overpowering feeling that something is "wrong", eliciting compulsive thoughts or behavioral patterns intended to somehow make it "right" (Schwartz and Begley 2002).

Bipolar or manic-depressive disorder (BD) is characterized by extreme mood swings between alternatively euphoric and depressed states. Cyclothymic disorder is a milder form of this illness (Goodwin and Jamison 1990). Several genetic regions have been implicated in these conditions, including six specific chromosomes (Craddock and Jones 1999; Craddock et al. 2005). Since Schildkraut (1965) suggested the involvement of the neurotransmitters norephinephrine and serotonin and with the advent of neuroimaging a number of brain areas have been implicated. This includes the observation of reduced gray matter in the left subgenual prefrontal cortex and amygdala enlargement (Vawter et al. 2000), and decreased neuronal and glial density in association with glial hypertrophy (Rajkowska 2009). Significant shape differences have also been observed in caudate and putamen, thus implicating the basal ganglia (Hwang et al. 2006).

Schizophrenia afflicts the frontal lobes with connectivity problems, contributing to the appearance of atrophy, and the cingulate cortex, temporal lobes, and hippocampus are all adversely affected. The illness involves volumetric changes of gray matter in the right and left middle and inferior temporal gyrus, worsening with chronicity (Kuroki et al. 2006; Hershfield et al. 2006). Hippocampal volume is also reduced in schizophrenia and there is lateral ventricular enlargement (Harrison 1999). The condition is associated with frontal lobe dysfunction and disconnectivity (Mathalon and Ford 2008). These morphometric changes are suggestive of alterations in synaptic, dendritic, and axonal organization, a view supported by immunocytochemical and ultrastructural findings. Pathology in subcortical structures is not well established apart from the dorsal thalamic nuclei, which are interconnected with the dorsolateral prefrontal cortex (which possesses VENs). The anterior cingulate cortex containing large numbers of VENs is also involved. Reductions in the number of small neurons in layer II and reduced cerebral blood flow in the anterior cingulate have been noted in schizophrenia (Tamminga et al. 1992).

Multiple sclerosis (MS) is apparently but not conclusively an autoimmune inflammatory disease of the central nervous system, causing demyelination of axons (Sailor et al. 2003). It is characterized by multifocal lesions, the MS plaques. Activated mononuclear cells destroy myelin and to some degree oligodendrocytes, the glial cells that produce the myelin in which axons are wrapped. Remaining oligodendrocytes attempt to produce new myelin, but in most cases this pattern of inflammatory reaction subsides only to appear at another location or at another time. The pattern of progression suggests a relative hierarchy of changes over time, involving first frontal and temporal regions (Lumsden 1970) and later the precentral gyrus (Wegner and Mathews 2003). MS can impact on any area in the central nervous system although visual areas are commonly affected; onset occurs most often in young adults but can even be before age 15 or after age 50 (Compston and Coles 2008).

Autism spectrum disorders manifest themselves in early childhood and their etiology remains unknown. Abnormalities have been detected in the frontal and temporal lobes, the cerebellum, the amygdala, and the hippocampus. VENs have been implicated in autism (Allman et al. 2005: 367) but this finding has been challenged (Kennedy et al. 2007). Underconnectivity in the brains of children with autism (Hughes et al. 1997) offers a basis for further investigation of this and other pervasive developmental disorders (Brasic 2009). In some subgroups, cerebellar dysfunction may occur, in others there is dysfunction of the prefrontal cortex and of connections to the parietal lobe. Reduced activation in the fusiform gyrus, the portion of the brain associated with facial recognition, and increased activation of adjacent portions of the brain associated with recognition of objects have been observed. The amygdalas of patients with autism have fewer nerve cells, especially in a subdivision called the lateral nucleus of the amygdala (Balter 2007).

Asperger's syndrome differs from autism in that it lacks the aberrations or delays in language development or cognitive development that are typical of autism. Sufferers may also have normal or even superior intelligence, in contrast to the low IQ associated with autism. However, they share social insensitivity and other characteristics with autism patients. Asperger's also develops in early childhood, generally after the age of three.

Temporal lobe epilepsy involves the limbic system and may originate in several locations within the temporal lobe, the hippocampus, parahippocampal gyrus, amygdala, etc. (Benson and Blumer 1975). This epileptic condition is included here because of its effects on the limbic system (Volle and Heron-Helvenston 1979), which it shares with many of the other pathologies listed.

Middle cerebral artery stroke, although not related to neurodegenerative conditions, is considered because it is the leading neurological illness, and evolution would have strongly selected against this susceptibility. Stroke derives from thromboemboli lodged in a cerebral blood vessel; from platelet emboli; or from carotid or vertebral stenosis. The middle cerebral artery is the largest cerebral artery and also the most commonly affected by cerebrovascular incidents. Since it supplies large areas of the brain (most of the outer convex brain surface, nearly all the basal ganglia, and the posterior and anterior internal capsules) infarcts can lead to diverse conditions. These include apraxia (inability to perform previously learned physical task) and dyspraxia (inability to perform a physical task), and Broca's and Wernicke's aphasia (expressive and receptive language deficits).

This brief list of brain disorders should suffice to show not only that specific brain regions are selectively affected by them, but also that these are in most cases those very same regions facilitating what are generally defined as the "higher cognitive functions" of the human brain (Damasio et al. 1990). There is every possible indication that this neurological susceptibility is directly linked to the complexity of the ever-burgeoning brain. Expressed in simplistic terms, it has given us both the genius of our greatest thinkers and artists, and the despair of "losing our mind". This immediately raises a fascinating question: did our ancestors during early parts of the Pleistocene suffer from the neurodegenerative curse? If the two extreme conditions of the human mind are in some measure correlated, the answer might be, these afflictions appeared only as hominin cognition became overly complex, perhaps as if it placed excessive demands on new neural structures, or if natural selection had failed to select against them. If we were to investigate the origins of mental illnesses this is surely where we would need to begin our inquiry. Until now, three closely related issues of explaining these matters have not been investigated in any depth: (1) how did these pathologies initially develop; (2) at what time in our evolution did they appear; and (3), most importantly, why did evolutionary processes apparently fail to select against the relevant genetic predispositions? Without some appreciation of these issues the diseases concerned have no causal context or explanation; we are merely trying to make sense of end effects without an appreciation of how they came about. Science, however, expects some level of causal reasoning from us. Here I will attempt providing some useful thoughts on this.

The Great Paradox of Recent Human Evolution

We have seen that the ventromedial or orbital prefrontal cortex has been implicated in human cognitive evolution as well as in the attendant pathologies, and the involvement of forebrain-neuroanatomy and neurophysiology of frontal and prefrontal structures in mental illness has been demonstrated (e.g., Heimer et al. 2008). This nexus between emerging advanced cognitive abilities and neurodegenerative susceptibility, therefore, requires detailed attention (Bednarik and Helvenston 2011). In some cases we already have first genetic indications that such predisposition may have been limited to "modern" humans. For instance, the DYRK1A gene, implicated in causing Down syndrome, seems absent in robust *Homo sapiens*. The genes CADPS2 and AUTS2, responsible for autism, also appear to be limited to modern Graciles. Perhaps more dramatic is the proposal that schizophrenia is of late historical origin and might have been introduced by a virus as recently as 200 or 300 years ago (Jeste et al. 1985; Hare 1988). Indeed, there appears to be no earlier mention of the disease, in contrast to other conditions such as bipolar illness (Bednarik and Helvenston 2011). Be that as it may, the NRG3 gene, associated with schizophrenia, also seems to be absent in so-called Neanderthals. Using the human haplotype map to test for selective sweeps in regions associated in genome scans with psychosis, such as 1q21, is promising (Voight et al. 2006). Again, such selective sweeps tend to yield relatively recent etiologies, of less than 20 ka, as predicted by the domestication hypothesis. While the genes SLC6A4 and NRG1 have been implicated in schizophrenia, and MAOA linked with bipolar disease (Cho et al. 2005; Li et al. 2006; Andres et al. 2004; Preisig et al. 2005; Jansson et al. 2005), few such loci have been identified for polygenic conditions (but see Saito et al. 2001; Yoshikawa et al. 2001; Ding et al. 2002; Enard et al. 2002; Muglia et al. 2002; Zhang et al. 2002; Kitano et al. 2004; Spinks et al. 2004; Stopkova et al. 2004; Abdolmaleky et al. 2005; Costas et al. 2005; Harrison and Weinberger 2005; Gardner et al. 2006; Sanjuan et al. 2006; Xu et al. 2006, which collectively link schizophrenia or bipolar disorder with APOL1, APOE, DRD4, FOXP2, GRM3, HOPA, IMPA2, MAOA, MAOB, NRG1, SLC6A4, and SYNJ1). This contrasts with the successful identification of more than 1700 Mendelian (single gene) disorders. Continuing research is likely to locate more evidence that neurodegenerative illnesses are the burden specifically of modern sapienoids, just as other "modern" human genes such as RUNX2 and CBRA1 (causing cleidocranial dysplasia or delayed closure of cranial sutures, malformed clavicles, and dental abnormalities) and THADA (associated with type 2 diabetes) are certainly deleterious. In a species fully subject to the canons of natural selection, such disadvantageous mutations would surely tend to be suppressed.

The greatest mystery in paleogenetics or evolutionary genetics is the conundrum of why evolutionary processes apparently failed to select against the degenerative genetic predispositions of extant humans. In the perhaps most comprehensive debate of this topic, Keller and Miller (2006) defined it as that field's unresolved paradox. Although their discussion and the contributions by 35 of the foremost specialists in the field truly considered every possible aspect and perspective of this veritable Gordian knot, it failed in providing more than a selection of weak and inconclusive hypotheses. Keller and Miller propose three alternative evolutionary models—ancestral neutrality, balancing selection and mutation–selection balance—but in the end fail to implicate any one of them. Why the maladaptive mental disorder susceptibility alleles have not either fixated (for which they would need to be adaptive) or gone extinct has, therefore, remained entirely unsolved so far. This book on *The Human Condition*, however, not only provides a realistic and logical explanation for this evolutionary puzzle, but also explains why it has for so long remained unsolved. By promoting the replacement hypothesis as aggressively as Pleistocene archeologists have done over recent decades, neuroscientists had to conduct their deliberations within a false paradigm, i.e., the notion that today's humans are a distinctive species and their characteristics are attributable to natural selection. Within that framework, the selection in favor of numerous deleterious traits is indeed an unsolvable paradox. Within my domestication hypothesis, by contrast, it is not only readily explained, but also entirely predictable and logical. Domestication promotes unfavorable alleles (e.g., Horrobin 1998, 2002; Andolfatto 2001; Lu et al. 2006), and it can even account for other unexplained features, such as the abolition of estrus in females (see Chapter 5), which is as typical of domestication as are so many detrimental traits.

The evolution of mental and cognitive faculties in hominins should be assumed to have also involved a gradual change from impulsive toward the obsessive range of a spectrum of behavior (cf. Ochse 1990; Bednarik 2011). The apparent absence of neurodegenerative and other mental illnesses in extant nonhuman primates seems to point in the same direction. We need to recall that, based on judicious archeological reasoning, a developing sense of perfection appears to have been developed by the Middle Pleistocene, and it can apparently be traced back to Mode 1 technocomplexes (see first part of Chapter 5). The neuroscientific information seems to imply that these ancestors, late *H. erectus* and early robust *H. sapiens*, had developed specific brain structures and tissues that facilitated advanced mental and cognitive performance. In an incipient form, Merlin Donald (1991) had already touched on the irony of "advancement" through relinquished control. As the mental faculties of hominins increased and hominins rose to the top of the food chain, individual reliance was delegated to society and to objects, the latter ranging from tools to objects of externally storing symbolic information. But to what extent were our ancestors at the same time rendered progressively more susceptible to neurodegenerative and neurological pathologies?

The concept of perfectionism, which we have also visited in Chapter 5, certainly has adaptive value in a cultural system, but it does require a level of obsessiveness. Obsessive-compulsive disorders seem to illustrate that obsessive and neurotic behavior is the price we pay for our rapid cognitive evolution; they are rather like an unwelcome side effect of it. The argument can be extended to most cultural behavior of humans, which, when viewed pragmatically, is frequently irrational. Seen in a realistic perspective, it is often overwrought and involves countless obsessive aspects, for example, our regimented behavioral routines or our relentless acquisition of excessive surpluses as an insurance against the loss of another insurance, which surely would qualify as an obsessive behavior symptom. Our access to "objective reality" (should such a state exist—which we remain ignorant about; see Chapter 1) is severely limited by the residues of irrational ideologies, baseless ontologies, and metaphysical straightjackets, the baggage of many millennia of cultural, including religious, development. Our hankering for something to believe in is not shared by other primates. Our reliance on "specialists" (shamans, prophets, scientists, medicos—or for that matter, writers addressing human origins) ignores that *déformation professionnelle* is also a distortion in the way the world is perceived. Professional training involves obsessive behavior traits that may be neurophysiologically relatable to compulsive behavior. Indeed, it is almost a tautology to say that without obsessions humans would not have reached the level of cultural sophistication they have.

If that is the case we may legitimately ask, at which stage in human development can evidence for these features first be detected? This book has provided such evidence. The unnecessarily perfect appearance and symmetry of many handaxes suggests to some researchers that the "conscious" quest for perfection may have begun to appear with the Acheulian tool traditions that created these forms (e.g., Gamble 1997; Wynn 1979, 1989, 2002). It was certainly well developed by the time ostrich eggshell beads or cupules were made (see Chapter 5). In a neurological sense this is an important marker because a sense of perfection is certainly an obsessive trait. In a neo-Darwinian sense it offers rather little benefit to us, yet it is highly time-consuming and demanding on our resources. One could argue that there is perfection in the construction of a spider's web or a bird's nest, but it is assumed that the specifications of these structures are genetically encoded; there is no "conscious" initiative involved. Moreover, there is no sense of perfection apparent in anything extant nonhuman primates make. Their sleeping nests or tools reveal no compulsion to go beyond the purely functional and have very probably remained completely unchanged for a long time: no desire to improve is evident. Nor is any such obsession apparent from the tools of australopithecines or early *Homo* (up to *H. erectus*). Evidence for these impulses seems to appear roughly midway through the Acheulian technocomplex, or close to a million years ago. This coincides with the introduction of seafaring and the first clear evidence for displacement-based language (*sensu*

Bickerton 2010, and see Chapter 4). Therefore, it seems reasonable, at least until contradicting evidence comes to light, to attribute the rise of exograms—of external storage of symbolic information—essentially to the Middle Pleistocene period. It is here that the origins of human modernity begin to become apparent on the empirical record.

By the same token, as that very process appears to be implicated also in the rise of neurodegenerative diseases, and in the almost countless other detrimental genetic disorders humans are saddled with, it needs to be assumed that these, too, developed during the last million years or so. This is clearly a disadvantageous development for humans, and we can only assume that, at any one stage, the advantages of the development of the prefrontal cortex must have significantly outweighed its detrimental effects. Which might suggest that the latter were negligible initially, otherwise they could have selected against the changes giving rise to cognitive sophistication. So at what stage should these drawbacks be assumed to have become significant phylogenetic encumbrances of the human genome? *And why did Darwinian evolution apparently fail to select against such unfavorable traits?*

This is a difficult question and perhaps not answerable by empirical observation. But if we consider the hypothesis presented in Chapter 5, attributing Final Pleistocene human neoteny to cultural selection, a realistic explanation offers itself. Just as evolutionary determinants could not prevent the deleterious changes from robust to gracile forms, because they were overruled by cultural determinants, the mental diseases arising from burgeoning prefrontal cortex complexity may have escaped natural selection in much the same way. The development and persistence of these diseases seems to signal that selection against them was somehow muted. It could also be taken as an indicator that the selective benefits of the Late Pleistocene development of cognition and symboling were of such outstanding effects that the "byproduct" of neuropathologies was tolerated—or rather, not significantly selected against. Perhaps the level of tolerance of these consequences rose in proportion to the selective benefits of developing cognition.

The principal effect of the rise of symbol-based culture, then, has been its facilitation of changes not dictated by natural evolution, but determined by cultural evolution. This is not to suggest natural selection has been abolished for humans; rather that cultural factors have begun to codetermine and eventually dominate changes to our genome. The gene-culture coevolutionary model we have visited in Chapter 5 predicts such a development, and some of the evidence and propositions presented in this book document it. The gracilization marking human development over the past fifty millennia was not an evolutionary process; it was determined by cultural selection. Selection for a single trait results in changes in numerous traits—changes that are usually deleterious, be they physiological or a range of others (see Chapter 5). The resulting loss of robusticity involved several reductions in evolutionary fitness, for example, the size of the brain decreased at a time when demands made of it are thought to have increased dramatically. This development was, however, offset by the increased use of exograms, which eventually reached such extraordinary proportions that today almost all of our symbolic information resides outside our brain. By the same token, as the rising specter of perfection and performance became a selection factor, it also eroded neurological balance. The developing neuroses could no longer be selected against, because instead of natural selection culling the relevant alleles, these had acquired *cultural selection* capital.

Assembling the Puzzle Correctly

The process of selecting genetic traits by means other than Darwinian evolution is domestication, a radical hereditary reorganization of the genetic constitution of a species. In the case of robust *Homo sapiens*, it significantly accelerated the fetalization of the species that had already begun earlier, promoting behavioral and other plasticity. Reflected in the change from nonfigurative symboling to figurative depiction, it is this plasticity that made the cognitive developments possible that mark the last few dozen millennia—and not the appearance of some mythological invaders (Bednarik 2008). The replacement explanation is now as redundant as the proposition that the primitives of the Pleistocene could never have made the artistic masterworks in the Franco-Cantabrian caves. Pleistocene archeology has made a great many discoveries since archeologists rudely denounced the very notion of humans in the Pleistocene; it has assembled an immense amount of jumbled data. But it has yet to master the art of assembling these jigsaw puzzle pieces correctly, because it has been dominated by a process of explanation that, among other things, tends to begrudge the ancients their achievements. It has, in a sense, participated in the age-old practice of defining societies by contrasting us with "The Other", which is consistent with the original charter of archeology—to invest the emerging nation states of the nineteenth century with origin myths. It has always been an intrinsically political discipline (Trigger 1984, 1985, 1989; Silberman 1995; Kohl and Fawcett 1995). Thus, the innate ardor to place the greatest possible cultural distance between our early ancestors and us has evolved a biased discipline. That bias has for the last 150 years sought to reject any evidence of sophistication of Pleistocene hominins. It has preferentially collected evidence of primitiveness and sought to emphasize this in its narratives. But to be politically acceptable, all extant human societies had to be included in this brotherhood of modern humans, irrespective of contradictory somatic data. History, on the other hand, operates on the principle of extolling the values of late Holocene cultural evolution and its grandiose achievements. These tensions have resulted in the construct of a pinnacle of evolution—present man—whose intelligence, compulsions, and obsessions drive his puerile search for something to believe in that stands in stark contrast to the innate maturity of all other species. Indeed, the entire purpose of evolution was apparently to produce this paragon of virtue, this likeness of a deity.

Needless to say, the proposition that we are a neotenous ape susceptible to numerous genetic disorders, a creature obsessed by perfection but itself genetically very imperfect, is not going to be very popular. Humans are also vain, and veracity is not the purpose of their anthropocentric humanism. Of relevance in this context is the inclination of the members of human societies to perceive their own cultures, at least subconsciously, as existing at some kind of developmental zenith. Just as political or ethnic entities typically define themselves by contrasting their virtues with the perceived deficiencies of others, each group tends to see itself as technologically and culturally sophisticated, usually at the expense of others. The practices of archeology, with its neo-colonialist tendencies toward those it studies—which in a sense justify its very existence—offer an opportunity of reifying such concepts. Throughout its history, archeology has suffered from underestimating the sophistication of the societies it has tried to study. Jean François Champollion, by deciphering the Egyptian hieroglyphs, gained an access to such cultural complexity it still seems incredible 180 years later—and some still today doubt the achievements of these Bronze Age people, looking instead for extraterrestrial explanations. When Don Marcelino Santiago Tomás Sanz de Sautuola reported the cave paintings of Altamira and implied that they were the work of Pleistocene people, he unleashed such a fury of archeologists that it led to his premature death. Such cultural complexity as that demanded by the cave paintings was simply incommensurable with the lowly tools of Stone Age brutes, and his challenge of the official dogma was heresy. Similar incredulity greeted my proposal, thirteen decades later, that some of the Paleolithic art was not even the work of people quite like us, but of "brutes" some archeologists would prefer to assign to the apes, those dreadful "Neanderthals" (Bednarik 2008). Worse still, I also advocated the view that hominins of the late part of the Early Pleistocene, some 800 or 900 millennia back, sailed the open sea to colonize numerous islands (Bednarik 1999).

Why such "blasphemous" postulates should be so repugnant to orthodox archeology, irrespective of their empirical basis, is an issue worthy of examination. Is it only because they contradict the dogma and thus challenge the discipline's power to explain the human past, or is there a deeper reason? These heretical ideas imply that the gap between the subject and the mediator's explanatory power might question archeological explanations in a generic sense. But this also confronts archeologists with an uncomfortable question: what is it that warrants their judgment of the level of complexity of cognitive frameworks of past societies? Unless they understand the rather significant epistemological shortcomings defining the reference frameworks of their own societies, as defined in Chapter 1, it would seem premature to judge the level of primitiveness of alien societies on the basis of their purported technologies. In fact there is something rather colonialist about a dominant society's self-appointed experts of the human past deciding such issues. The history of archeology shows unambiguously that its practitioners failed consistently in detecting the level of nontechnological sophistication of the societies whose material residues they studied, and that most narratives provided by Pleistocene archeology were ultimately false (Campbell 2006; cf. Searle 1995; Berger and Luckman 1966).

Apart from the inherent neo-colonialist nature of the discipline, this is to be expected also because, for the greater part, archeology deals merely with discarded materials—with the refuse of past cultures. This must surely provide an inadequate measure of sophistication at the best of times: as noted in Chapter 4, the contents of an American garbage disposal site will in a few millennia offer no evidence that the society in question was capable of placing humans on the Moon. To overcome this massive inherent bias, I began focusing on evidence for the supreme technological or cognitive capabilities of early societies—not just to obtain a more balanced perspective, but primarily to explore their limits. For instance, the pre-Historic equivalent of space travel is sea travel: numerous islands and at least one continent were colonized via the oceans tens or even hundreds of millennia ago. These quests, I reasoned, would have been cutting-edge achievements, barely possible at the time, and no doubt matters of life or death for the participants. Their relative degrees of difficulty, measured by distances traveled and adverse conditions overcome to succeed, can thus be assumed to provide precise information about the ultimate technological and cultural limits of the people concerned.

This seems such a reasonable strategy that I find it hard to understand the motivation of my most ardent opponents, who will go to any extreme to explain away evidence of early seafaring, or of equally early use of beads, or of Middle Pleistocene paleoart generally. For instance, they will insist that all the early maritime colonizations were accidental, with whole founding populations (of dozens of people!) each and every time drifting on floating vegetation mats, or floating on elephants. My question is not so much, how do they explain why of all the large land mammals, only humans and elephants crossed; or how they account for the numerous crossings of sea-narrows, none of which can be crossed without propelling energy. I am more interested in why it is that archeologists feel obliged to deny early humans certain levels of development—why they apparently regard any suggestion of early advanced human abilities as an assault on their authority. This, after all, has been their modus operandi for about 180 years. It would seem so much more sensible to start out with a null hypothesis, rather than a position that needs to be defended, a position denying early people the abilities of innovation, initiative, symboling, language, and so forth. In particular, there is no logical or obvious reason why archeologists could not have chosen the precise opposite position, allowing early hominins all reasonable capabilities and then gradually deleting those that were evidently absent. My impression is that the reason for the extreme position of Pleistocene archeologists is a subconscious religious conviction that the greatest possible intellectual, cognitive, and cultural distance should be maintained between "us" and those savage ancestors, because it is that distance that seemingly justifies the existence of archeology and certain religious beliefs. The African Eve Hoax is perhaps the most prominent expression of this need, contrasting the stupid Neanderthals with the smart Moderns. Never mind that the former had a significantly larger brain, were no doubt more intelligent and managed to preserve themselves for hundreds of millennia, whereas we are engaged in relentless self-destruction, driven by irrational ideologies and pathological insecurities deriving from our neurotic minds (Badcock 1980).

I find this completely illogical, because we have no grasp of how primitive our own conceptions of reality are, so how can we point at some other hominins and claim that theirs were vastly inferior? We modern humans of a more or less Western disposition (pinnacles of evolution, evidently) possess no access to objective reality; we do have constructs of reality but *we know that they are false* (and have known this since Plato's simile of the cave). Our entire conceptual framework (see

Chapter 1) is so conditioned by religion, culture, and other subjective factors that we would lack any ability of attaining real objectivity even if it were accessible to us. Consider by contrast the more mature Aborigines who, despite having to put up with our proselytizing, knew that humans are animals long before Darwin told us so. There are numerous other examples showing that Aborigines (and many other traditional societies) possessed ontologies that were superior to the beliefs held by Europeans until very recently. For instance, the Aboriginal nations explained many natural phenomena correctly since time immemorial. Having observed such features as volcanic flow patterns and fossil casts they believed that the rocks were once soft; if that had been proposed in Europe just a few centuries ago, it would have been religious heresy: the world was exactly as God created it. Surely a society that believes in a Santa Claus figure and an afterlife, and that our lineage is destined to tame nature and convert the heathens, is conceptually more primitive (and more damaging to the planet) than one that appreciates that humans derive from other animals; that they are part of nature; and that seeking to destroy nature will extinguish the basis of their existence. To judge the capabilities of our Pleistocene forebears by rejecting any evidence contradicting our preconceived dogma simply has no logical basis.

At the very beginning of this book I suggested that Darwin's greatest achievement was not his magnificent idea, but the fact that he was capable of arriving at it despite the ideological limits of the reality constructs society, education, and religion had imposed on him. It is almost impossible for us to fully appreciate the power of the conceptual systems within which we happen to exist—let alone resist it. These systems reflect the social needs and aspirations not only of the individual, but also of groups, classes or whole cultures—and they may be far from readily perceptible to us, who depend on them as our frames of reference.

The most dramatic changes marking the transition from robust to gracile *Homo sapiens* are perhaps not the somatic variations wrought by self-domestication, but the burgeoning growth in exograms and the reorganization of the neural structures connecting us to them. This, it appears, has had some "side effects", such as the appearance of a variety of brain disorders, such as those I listed above. But if it were correct that cultural preferences in mate choice began developing in the final Pleistocene, other details could also be speculated about. For instance, if there was a preference for females of young appearance, this might imply the forming of longer term relationships, because otherwise it would make no sense for a male to prefer a partner with a potentially longer stretch of fertile years. Similarly, if a pregnant or nursing female was provisioned with protein and lipid-rich food, it was more likely by one specific male rather than the group. But while we can safely assume, on the basis of empirical evidence, that there was a preference for neoteny, initially of females, resulting in domestication, any derivative hypotheses may be too precipitate.

Nevertheless, there are apparent indications that social structures had become rather complex by the period defined as the Early Upper Paleolithic. If one of the core propositions of this book is true—that the cultural sophistication of the ancients tends to be greatly underestimated—social constructs could be assumed to have been no less complex than those of extant hunter-fisher-foragers, such as the Australian Aborigines prior to colonization. For instance, if it were correct that much of the paleoart of the Aurignacian and Gravettian tool complexes is the work of teenagers, the notable sexual content of it would be readily relatable to oversexed male pubescents. These can only be the product of a social system barring them from ready sexual access to females as, for instance, in Aboriginal societies. If the content of the early iconographic imagery were any indication of its producers' preoccupation, which seems quite likely (Bednarik 1986), the imagery of dangerous animals, prey species and human female sexuality stand out (Bednarik 2007). The most economical interpretation of this would be that young males had to prove their courage and hunting prowess to gain the access they craved, so they were powerfully preoccupied with these subjects. The compulsions to have food, social acceptance and access to females are rather strong and elementary imperatives. I have spent a lifetime trying to understand rock art, and I have tried seeing it through the eyes of my teachers, who were "men of high degree" (i.e., Aboriginal elders; Elkin 1977); and as a scientist I tend to prefer simple explanations to intricate ones (Occam's razor). This is a very simple interpretation, but its validity cannot be demonstrated. It would signify that those adolescents of 25 or 35 ka ago lived in structured groups under the control of elders.

Becoming Human

In this book I have presented explanations for the process of becoming modern humans that differ so dramatically from those offered previously that they demand a radical revision of our theories about the course of our species' ascent. The perhaps most consequential, if not revolutionary, aspect of the version I favor is that what is traditionally seen as human evolution is only partly so; in its most recent phase, human development is increasingly moderated by unintended self-domestication. Throughout most of the Pliocene, the genesis of our lineage was determined largely or entirely by Darwinian evolutionary variables, such as those provided by the environment. However, in the late Pliocene and over the course of the subsequent Pleistocene, the exponential rise in the complexity of hominin culture began to affect humans in several ways. This proposition stands in stark contrast to the orthodox view, which perceives very little change for the entire duration of the so-called Lower Paleolithic, exceeding in duration two million years. It sees the cultures of this period as technologically almost static. However, at some point between 1.8 and 1.6 million years ago, the Mode 1 stone tools of Oldowan type gave way to the Acheulian traditions, marked by standardized artifact templates, especially the "handaxes". With time, these took on progressively more perfect forms, apparently well beyond the purely utilitarian. Therefore, I find it wrong to assume cultural stasis for this huge time span. Moreover, economy in stone tool design may be a sophisticated adaptation to high mobility (as we know especially from Australian ethnography), and apparent long-term conservatism in lithic typology can equally well indicate a succession of various cultures of a uniformly well-adapted stone tool technology.

Of more momentous consequences was the next wave of cultural metastasis, occurring roughly half way through the Pleistocene (i.e., almost one million years ago). It is marked by such achievements as seafaring colonizations, at least in two regions on the planet, and the development of taxonomic abstracts, such as the recognition and contemplation of exotic objects. Reflective language capable of displacement can be safely assumed to have been in use at that stage. Again, I stand in sharp opposition to the short-range archeologists, who see these developments as much more recent phenomena, attributable to the Late Pleistocene, and preferably to its most recent part. I am, however, in agreement with many scientists on this.

Next, my heretical model places the proliferating use of exograms, or evidence of symbol use, in the Middle Pleistocene, several hundred millennia into the past. My opponents reject this proposal completely, advocating the much later advent of such features, essentially in the last third of the Late Pleistocene (i.e., less than 50 ka ago).

But where I diverge from the traditional dogma of Pleistocene archeology most severely is in my explanation of the dramatic changes to human physiology as well as culture during these last 50 ka. The dominant theory is that a culturally superior species arose mysteriously in sub-Saharan Africa 200 ka ago, left that continent at least 100 ka ago, reached Australia 50 or 60 ka ago, and Europe perhaps 35 ka ago. It out-competed or extinguished all other humans of the world and was unable to interbreed with them. Therefore, no other human genes younger than 200 ka could have survived in these African super-humans. In recent years, a number of cracks have begun to appear in this model, but it still remains the discipline's favorite.

I perceive no evidence for this model, be it archeological, paleoanthropological, genetic or cultural (Bednarik 2008). In particular, I find the notion of tribes wandering through unpopulated landscapes absurd. We know that robust groups, probably Neanderthal-like people, lived in extremely hostile regions of the Old World, including in the Arctic, and we know that robust hominins settled numerous islands and one continent, presumably due to population pressures. People subsisted wherever the environment permitted human occupation, with the help of fire, shelters, and fur clothing. They even managed to survive temperatures below –40 or –50°C, which renders the "empty landscapes" notion hopelessly out of touch. We can safely assume that all reasonably hospitable parts of the Old World—and even some not so hospitable ones—were fully occupied by tribes for hundreds of millennia, possibly as far back as a million years ago. Wherever these people lived, they can be assumed to have been very well adjusted to their particular environmental niches, in every possible way.

In this scenario it is demographically naive to assume that a tribe of presumably naked Africans would have overwhelmed tribe after tribe in cold Eurasia, until there were no other people left. Not only would these invaders have lacked the technologies to subsist in cold climes, they would have fallen significantly short of the adaptive alleles the resident tribes had at their disposal. It is particularly important to repeat, as noted in Chapter 2, that in all cases where robust and gracile groups coexisted in the same region—in Europe, the Levant, Australia, in Africa itself—they had virtually identical economies and technologies, even ornaments. Nowhere do we have evidence that Graciles were in any sense superior to contemporary Robusts, and nowhere has it been shown that they were genetically very different. Comparing the genome of present-day humans with those of the Pleistocene Robusts, as is often done, is a vacuous exercise, because it tells us nothing new: we would not expect them to be identical. We would need to compare the genomes of contemporary populations. The much more important question to be asked is the core issue I have raised in this volume: why did the change from robust hominins to gracile ones occur? Why was it possible that deleterious developments took place, such as the burgeoning of mental diseases and thousands of other genetic disorders, the significant reduction in brain size, as well as the loss of cranial and other skeletal robusticity, coinciding with a similar diminishment of muscle strength? And why did these detrimental changes occur throughout the settled part of the world, and within roughly the same time frame?

These are the questions that needed to be asked, based on empirical evidence that can hardly be disputed. As we have seen, the answer of the replacement advocates is that these changes were very sudden, indeed instantaneous, because all existing populations were replaced by a new species. This hypothesis, I have suggested, has now about as much credibility as the idea that Australia's Wandjinas depict aliens from outer space. It joins the long list of questionable notions of archeology over the past couple of centuries. But this does not answer the question posed: how do we explain the global changes from Robusts to Graciles realistically? Here I have offered an outline of a realistic explanation. It may not be the right one, but it does have the advantage of accounting for all the empirical knowledge we currently have about this general theme. We may just have to reject the contrived extravagancies and hyperboles presented by the replacement apostles, beginning with those made by Protsch from 1973 onward (see Chapter 1), and followed by those of so many others I would not even attempt to list them all here.

My explanation, on the other hand, derives simply from the sound evidence we currently have. Gracilization occurred gradually in most regions and is, in fact, very probably still continuing today. The exponential development of culture and technology, marking the human ascent of the last fifty or so millennia, is closely linked to man's somatic development as recorded in skeletal remains. Indeed, it is in the final analysis culture that began to determine which genes would be replicated more often than others. In a small measure, that process may have commenced much earlier, but between 50 and 30 ka ago, a "critical juncture" was reached, in the sense that some people began being influenced by cultural constructs in selecting mating partners. For this to have a domesticating effect on populations, the selecting factor only needed to be a fraction of a percent per generation. After a few hundred generations, and with selective breeding apparently accelerating through time, significant changes occurred since that critical juncture. We have no need to speculate what these selected attributes were; they are readily documented in the changes we observe. Gracile characteristics became "fashionable", and these traits replaced those we define as robust.

Many sciences and "social sciences" may be interested in the processes that led to the modernity of hominins, be it in the somatic, cognitive, intellectual or cultural sense. These might include genetics, brain sciences, cognitive sciences, social neuroscience, paleoanthropology, behavioral sciences, sociology, and primatology. These sciences have to rely on Pleistocene archeology for information of the empirical evidence that might allow glimpses of specific manifestations of these processes, and for hypotheses about the course they could have taken. It appears that the instruction obtainable from archeology comprises much data tainted by indoctrination, by preconceived notions of what the data should show, and by the dogma established by the discipline's silverbacks. I am well aware that Pleistocene archeologists will, under the guise of testing them, move heavens and earth to disprove my propositions, but I am encouraged by the prospect that, once the sciences become aware that they have been subjected to archeological hoaxes, they are likely to react swiftly.

The Human Condition

In the first part of the first chapter we have considered the conceptual gaps between the different disciplines of the academic effort, how their practitioners "think differently, speak a different language, live in a different world" (Polanyi 1958). Language is heavily implicated in the formation of social reality and that of constructs of reality generally. The undeniable incommensurability of languages found at a small scale in academia also applies to entire societies and to humanity as a whole. Sapir's (1929: 209) profound dictum that "[n]o two languages are ever sufficiently similar to be considered as representing the same social reality" remains supremely relevant. No two humans live in precisely the same reality; each of us constructs their own reality over the course of their lives, and no two lives are identical. As Wittgenstein (1961) puts it, "[t]hat the world is *my* world shows itself in the fact that the limits of the language (*the* language which I understand) mean the limits of *my* world" (*Tractatus* 5.62).

Sapir continues, "[t]he worlds in which different societies live are distinct worlds, not merely the same worlds with different labels attached"—which brings us to archeology, a discipline where the term *déformation professionnelle* is particularly relevant. This is a discipline obsessed with attaching labels: to artifacts, human fossil remains, "cultural layers", perceived cultures, or to the carriers of these invented constructs (societies, nations, ethnic groups, perceived language or cultural groups, etc.). Nobody seriously believes in the emic currency of these designations, and yet these labels are used as if they had real, objective existence; as if they were real, definable cultures, societies or object types.

Here I have tried to set forth an all-embracing explanation of the human condition that defies archeological paradigms. That explanation gels well with quite a number of views but in a holistic sense it is at considerable variance with our dominant models of how or why we became the way we are, or even about the way we are. The human characteristics that have emerged as the perhaps most significant are certainly an odd assortment, differing as much from the idealistic views of yesteryear as day from night. In the final analysis it seems that the issue of a futile search for perfection is a key underlying factor in our becoming what we are. As individuals we are, each and everyone of us, imperfect, in the sense that we fall significantly short of our ideals. Our lives are ruled by obsessions, such as the search for something to believe in, or the obsession of unattainable perfection, all of which are already forms of obsessive behavior: chimps or bonobos could not be bothered with such nonsense. For us, our very existence revolves around them. Nearly all of modern human behavior is in the end neurotic, and modern society has turned being mentally unbalanced into a fine art, with a great many gradations and sophistries.

In particular, throughout the known history of man we see an obsession with exalting humans, be it by defining them as likenesses of gods or by glorifying human attributes, for example, in the arts. All of this chronic self-flattery, surely, is a little depressing for the individual, who cannot help but notice that he or she falls rather short of any ideal, in whatever sense. Every human being who has ever existed was a picture of imperfection, and apart from the odd psychopath, the rest of us could not help noticing this.

The question, then, is, can we cope with the truth about ourselves, or will we continue to lie to ourselves. Some people, the optimists, prefer the latter, especially those who are religious, have political or economic stakes, or prefer to uphold the status quo. Others, pessimists, hold that humans have, especially over the duration of the Holocene period, been more damaging than any other species in the history of this planet, and that our species seems to be afflicted with the seeds of self-destruction. Previous species and subspecies of humans have managed to share their environment sustainably with most living things for hundreds of millennia, whereas the recently emerged, neurotic *Homo sapiens sapiens* has fueled his wars against nature and himself with the resources of the world. For instance, vast forests were laid to waste to build tens of thousands of ships simply to ritually sink with cannon fire. Man has become the ultimate niche constructor—but there is also such a thing as negative niche construction, which occurs when a species exhausts the supporting capacity of a niche, or when the byproducts of its construction extinguish its advantages. Being trapped in its niche it is then destined for extinction.

Which raises the specter of humanity's future. It may be outside the scope of this volume, which as the title states is only intended to deal with the key questions of how we became what we are. But there can be no doubt that the propositions laid before the reader are also harbingers of considerable predictive potential. If the principal hypotheses of this book resemble what happened in the recent human past, and if the trajectories of such developments are extrapolated into the future, what should we expect? For instance, the process of neotenization should not be assumed to stop anytime soon, and if recent history were any guide, it could well be accelerating. Within a few generations we have in the Western world witnessed an unprecedented trend of young people increasingly shunning the responsibilities of maturity, and a strongly developing "cult of youth". This coincides with a probable increase in the incidence of neurodegenerative diseases, which may in part be attributable to greater life expectancies, but is perhaps also a result of long-term breeding trends. Then there is the cultural selection of compliant personalities over the past several millennia, because independently minded, individualistic or especially gifted or enterprising people have been selectively persecuted, executed, exiled or burned at the stake. This can have only exacerbated the effects of domestication, in which characteristics of compliance are usually selected for. It has also facilitated our latent (and sometimes fully developed) caste systems of social organization that tend to resemble those of social insects. As Bickerton (2010) points out in his magisterial investigation of language origins, ants too must have been free-roving organisms at some time in their evolution, so the comparison is not quite as far-fetched as it may seem. Seen in that perspective, many twenty-first century human societies have developed caste systems (e.g., of academic or economic apartheid) resembling those of ants, and while its members may well find this state reassuring and desirable, it is not at all clear what the long-term genetic effects of contemporary human breeding patterns will be. The modern welfare state has rendered most members of many societies unable to exist without such support, and has constructed new niches for humans, the genetic effects of which can be expected to be profound. They will, however, lie well outside of natural selection, and will profoundly accelerate processes begun several tens of millennia ago.

But where they will lead is certainly another story. Here we are more concerned with defining the human condition as it stands. If, as Carl Jung says, "the sole purpose of human existence is to kindle a light in the darkness of mere being", it is perhaps reasonable to expect human beings to learn to appreciate that they are not created in the image of a deity; far from it. They are neotenous apes whose brains have grown too fast, who have little or no comprehension of the reality of the universe or their place in it. They are the wrath of this planet (its skin disease, as Nietzsche would have it), and the principal justification for their continued existence seems to be that once in a while they are capable of producing genius. If we consider the true potential of this species, there can be no doubt that humans have fallen immensely short of what they could have been. This disappointing performance is in the end attributable to our defective brain, which we use for purposes it was never "designed," for example, to map reality. The notion that we are the arbiters of defining reality is preposterous, as pointed out by Plato some 2400 years ago, yet individually we remain addicted to it to the point of happily sacrificing our lives to the mirages it conjures up. The human brain is not just an organ susceptible to failing more often than any other; it is inherently a faulty organ.

References

Abdolmaleky, H. M., S. Thiagalingam and M. Wilcox 2005. Genetics and epigenetics in major psychiatric disorders: dilemmas, achievements, applications, and future scope. *American Journal of Pharmacogenomics* 5: 149–160.

Adolphs, R., D. Tranel, H. Damasio and A. R. Damasio 1994. Impaired recognition of emotion in facial expression following bilateral damage in the human amygdala. *Nature* 372: 669–672.

Adolphs, R., D. Tranel, S. Hamann, A. W. Young, A. J. Calder, E. A. Phelps, A. Anderson, G. P. Lee and A. R. Damasio 1999. Recognition of facial emotion in nine subjects with bilateral amygdala damage. *Neuropsychologia* 37: 1111–1117.

Allman, J., A. Hakeem, J. M. Erwin, E. Nimchinsky and P. Hof 2001. The anterior cingulate cortex: the evolution of an interface between emotion and cognition. *Annals of the New York Academy of Science* 935: 107–117.

Allman, J., A. Hakeem and K. Watson 2002. Two phylogenetic specializations in the human brain. *Neuroscientist* 4: 335–345.

Allman, J. M., K. K. Watson, N. A. Tetreault and A. Y. Hakeem 2005. Intuition and autism: a possible role for von Economo neurons. *Trends in Cognitive Science* 9(8): 367–373.

Andolfatto, P. 2001. Adaptive hitchhiking effects on genome variability. *Current Opinion in Genetics and Development* 11: 635–641.

Andres, A. M., M. Soldevila, A. Navarro, K. K. Kidd, B. Oliva and J. Bertranpetit 2004. Positive selection in MAOA gene is human exclusive: determination of the putative amino acid change selected in the human lineage. *Human Genetics* 115: 377–386.

Badcock, C. R. 1980. *The psychoanalysis of culture*. Basil Blackwell, Oxford.

Bailey, J. 2006. A brief overview of chimpanzees and aging research. Written for project R & R: Release and restitution for chimpanzees in US Laboratories. Internet address www.releasechimps.org

Balter, M. 2007. A mind for sociability. *ScienceNow Daily News* July 27: 1.

Barres, B. A. 2008. The mystery and magic of glia: a perspective on their roles in health and disease. *Neuron* 60(3): 430–440.

Bednarik R. G. 1986. Comment on W. Davis, 'The origins of image making'. *Current Anthropology* 27: 202–203.

Bednarik, R. G. 1999. Maritime navigation in the Lower and Middle Palaeolithic. *Comptes Rendus de l'Académie des Sciences Paris, Earth and Planetary Sciences* 328: 559–563.

Bednarik, R. G. 2007. Antiquity and authorship of the Chauvet rock art. *Rock Art Research* 24: 21–34.

Bednarik, R. G. 2008. The mythical moderns. *Journal of World Prehistory* 21(2): 85–102.

Bednarik, R. G. 2011. Genetic drift in recent human evolution? In F. Columbus (ed), *Genetic drift*. Nova Science Publishers, New York, NY.

Bednarik, R. G. and P. A. Helvenston 2011. The nexus between neurodegeneration and advanced cognitive abilities. *Anthropos* 106(2).

Benson, D. F. and D. Blumer 1975. *Psychiatric aspects of neurologic disease*. Grune and Stratton, New York, NY.

Berger, P. L. and T. Luckmann 1966. *The social construction of reality: a treatise in the sociology of knowledge*. Anchor Books, Garden City, NY.

Bickerton, D. 2010. *Adam's tongue: how humans made language, how language made humans*. Hill and Wang, New York, NY.

Blinkov, S. M. and I. I. Glezer 1968. *The human brain in figures and tables: a quantitative handbook*. Basic Books, Inc., Plenum Press, New York, NY.

Brasic, J. R. 2009. Autism. E Medicine Medscape. Internet address: http://emedicine.medscape.com/article/912781-print

Brodmann, K. 1912. Neue Ergebnisse über die vergleichende histologische Lokalisation der Grosshirnrinde mit besonderer Berücksichtigung des Stirnhirns. *Anatomy Anzeiger* (Suppl) 41: 157–216.

Brown, M. M., C. Kaplan, L. A. Jason and C. B. Keys 2010. Subgroups of chronic fatigue syndrome based on psychiatric disorder onset and current psychiatric status. *Health* 2(2): 90–96.

Butti, C., C. C. Sherwood, A. Y. Hakeem and J. M. Allman 2009. Total number and volume of von Economo neurons in the cerebral cortex of cetaceans. *Journal of Comparative Neurology* 515(2): 243–259.

Buxhoeveden, D. and M. F. Casanova 2002. The minicolumn and evolution of the brain. *Brain, Behaviour and Evolution* 60: 125–151.

Buxhoeveden, D., W. Lefkowitz, P. Loats and E. Armstrong 1996. The linear organization of cell columns in human and nonhuman anthropoid Tpt cortex. *Anatomy and Embryology* 194(1): 23–36.

Cáceres, M., C. Suwyn, M. Maddox, J. W. Thomas and T. M. Preuss 2007. Increased cortical expression of two synaptogenic thrombospondins in human brain evolution. *Cerebral Cortex* 17: 2312–2321.

Campbell, F. 2006. Molesting the past. *The Weekend Australian*, 25 February: R15.

Cho, H. J., I. Meira-Lima, Q. Cordeiro, L. Michelon, P. C. Sham, H. Vallada and D. A. Collier 2005. Population-based and family-based studies on the serotonin transporter gene polymorphisms and bipolar disorder: a systematic review and meta-analysis. *Molecular Psychiatry* 10: 771–781.

Christopherson, K., E. Ullian, C. Stokes, C. Mullowney, J. Hill, A. Agah, J. Lawler, D. Mosher, P. Brownstein and B. Barres 2005. Thrombospondins are astrocyte-secreted proteins that promote CNS synaptogenesis. *Cell* 120: 421–433.

Coghlan, A. 2006. Whales boast the brain cells that 'make us human.' *New Scientist*, http://www.newscientist.com/article.ns?id=dn10661&print=tru.

Compston, A. and A. Coles 2008. Multiple sclerosis. *Lancet* 372(9648): 1502–1517.

Costas, J., A. Salas, C. Phillips and A. Carracedo 2005. Human genome-wide screen of haplotype-like blocks of reduced diversity. *Gene* 349: 219–225.

Craddock, N. and I. Jones 1999. Genetics of bipolar disorder: review article. *Journal of Medical Genetics* 26: 585–594.

Craddock, N., M. C. O'Donovan and M. J. Owen 2005. The genetics of schizophrenia and bipolar disorder: dissecting psychosis. *Journal of Medical Genetics* 42: 193–204.

Crow, T. J. 1997. Is schizophrenia the price that *Homo sapiens* pays for language? *Schizophrenia Research* 28: 127–141.

Damasio, A. R. 1985. The frontal lobes. In K. Heilman and H. Valenstein (eds), *Clinical neuropsychology*, pp. 339–375. Oxford University Press, Oxford.

Damasio, A. R., D. Tranel and H. Damasio 1990. Individuals with sociopathic behavior caused by frontal damage fail to respond autonomically to social stimuli. *Behavioural Brain Research* 41: 81–94.

Diller, K. C., W. A. Gilbert and T. D. Kocher 2002. Selective sweeps in the human genome: a starting point for identifying genetic differences between modern humans and chimpanzees. *Molecular Biology and Evolution* 19: 2342–2345.

Ding, Y. C., H. C. Chi, D. L. Grady, A. Morishima, J. R. Kidd, K. K. Kidd, P. Flodman, M. A. Spence, S. Schuck, J. M. Swanson, Y. P. Zhang and R. K. Moyzis 2002. Evidence of positive selection acting at the human dopamine receptor D4 gene locus. *Proceedings of the National Academy of Sciences of the United States of America* 99(1): 309–314.

Donald, M. 1991. *Origins of the modern mind: three stages in the evolution of culture and cognition*. Harvard University Press, Cambridge, MA.

Elkin, A. P. 1977. *Aboriginal men of high degree: initiation and sorcery in the world's oldest tradition*. University of Queensland Press, St Lucia.

Enard, W., M. Przeworski, S. E. Fisher, C. S. Lai, V. Wiebe, T. Kitano, A. P. Monaco and S. Pääbo 2002. Molecular evolution of FOXP2, a gene involved in speech and language. *Nature* 418: 869–872.

Frank, L. M., E. N. Brown and M. A. Wilson 2000. Trajectory encoding in the hippocampus and entorhinal cortex. *Neuron* 27: 169–178.

Gamble, C. 1997. *Handaxes and Palaeolithic individuals*. In N. Ashton, F. Healey and P. Pettitt (eds), *Stone Age archaeology*, pp. 105–109. Monograph 102, Oxbow Books, Oxford.

Gannon, P. J., R. L. Holloway, D. C. Broadfield and A. R. Braun 1998. Asymmetry of chimpanzee planum temporale: humanlike pattern of Wernicke's brain language area homolog. *Science* 279: 220–222.

Gannon, P. J., N. M. Kheck and P. R. Hof 2001. Language areas of the hominoid brain: a dynamic communicative shift on the upper east side planum. In D. Falk and K. R. Gibson (eds), *Evolutionary anatomy of the primate cerebral cortex*, pp. 216–240. Cambridge University Press, New York, NY.

Gardner, M., A. Gonzalez-Neira, O. Lao, F. Calafell, J. Bertranpetit and D. Comas 2006. Extreme population differences across Neuregulin 1 gene, with implications for association studies. *Molecular Psychiatry* 11: 66–75.

Geschwind, N. and W. Levitsky 1968. Human brain: left-right asymmetries in temporal speech region. *Science* 161: 186–187.

Gloor, P. 1990. Experiential phenomena of temporal lobe epilepsy. *Brain* 113: 1673–1694.

Gloor, P. 1992. Role of the amygdala in temporal lobe epilepsy. In A. P. Aggleton (ed), *The Amygdala*. Wiley-Liss Press, New York, NY.

Goodwin, F. K. and K. R. Jamison 1990. *Manic depressive illness*. Oxford University Press, Oxford.

Hakeem, A. Y., C. C. Sherwood, C. J. Bonar, C. Butti, P. R. Hof and J. M. Allman 2009. Von Economo neurons in the elephant brain. *Anatomical Record* 292(2): 242–248.

Hare, E. H. 1988. Schizophrenia as a recent disease. *British Journal of Psychiatry* 153: 521–531.

Harrison, P. J. 1999. The neuropathology of schizophrenia: a critical review of the data and their interpretation. *Brain* 122: 593–624.

Harrison, P. J. and D. R. Weinberger 2005. Schizophrenia genes, gene expression, and neuropathology: on the matter of their convergence. *Molecular Psychiatry* 10: 40–68.

Hayashi, M. 2006. Spindle neurons in the anterior cingulate cortex of humans and great apes. In T. Matsuzawa, M. Tomonaga and M. Tanaka (eds), *Cognitive development in chimps*, pp. 64–74. Springer, Tokyo.

Heimer, L., G. W. van Hoesen, M. Trimble and D. S. Azhm 2008. *Anatomy of neuropsychiatry*. Elsevier, Boston, MA.

Hershfield, A., D. Yurgelun-Todd, R. Kikinis, F. A. Jolesz and R. W. McCarley 2006. Middle and inferior temporal gyrus gray matter volume abnormalities in first-episode schizophrenia: an MRI study. *American Journal of Psychiatry* 163(12): 2103–2110.

Hodgson, D. and P. A. Helvenston 2006. The emergence of the representation of animals in paleoart: insights from evolution and the cognitive, limbic and visual systems of the human brain. *Rock Art Research* 23: 3–40.

Hof, P. R. and E. Van der Gucht 2007. Structure of the cerebral cortex of the humpback whale. *Megaptera novaeangliae* (Cetacea, Mysticeti, Balaenopteridae). *Anatomical Record* 290(1): 1–331.

Holloway, R. L. 1995. Toward a synthetic theory of human brain evolution. In J. P. Changeaux and J. Chavillon (eds), *Origins of the human brain*, pp. 42–55. Clarendon Press, Oxford.

Holloway, R. L. 1996. Evolution of the human brain. Ch. 4 in A. Lock and C. Peters (eds), *Handbook of human symbolic evolution*, pp. 74–116. Clarendon Press, Oxford.

Holloway, R. L. 2001. Revisiting australopithecine visual striate cortex: newer data from chimpanzee and human brains suggest it could have been reduced during australopithecine times. In D. Falk and K. R. Gibson (eds), *Evolutionary anatomy of the primate cerebral cortex*, pp. 177–186. Cambridge University Press, New York, NY.

Horrobin, D. F. 1998. Schizophrenia: the illness that made us human. *Medical Hypotheses* 50: 269–288.

Horrobin, D. 2002. *The madness of Adam and Eve: how schizophrenia shaped humanity*. Bantam, London.

Hughes, C., I. Soares-Boucaud, J. Hochmann and U. Frith 1997. Social behaviour in pervasive developmental disorders: effects of informant, group and 'theory-of-mind'. *European Child and Adolescent Psychiatry* 6(4): 191–198.

Hwang, J., I. K. Lyoo, S. R. Dager, S. D. Friedman, J. S. Oh, J. Y. Lee, S. J. Kim, D. L. Dunner and P. F. Renshaw 2006. Basal ganglia shape alterations in bipolar disorder. *American Journal of Psychiatry* 163(2): 276–285.

Hyman, B. T., G. W. VanHoesen, A. R. Damasio and C. L. Barnes 1984. Alzheimer's disease: cell-specific pathology isolates the hippocampal formation. *Science* 225(9): 1168–1170.

Jansson, M., S. McCarthy, P. F. Sullivan, P. Dickman, B. Andersson, L. Oreland, M. Schalling and N. L. Pedersen 2005. MAOA haplotypes associated with thrombocyte-MAO activity. *BMC Genetics* 6: 46.

Jerison, H. J. 1973. *Evolution of the brain and intelligence*. Academic, New York, NY.
Jeste, D. V., R. Del Carmen, J. B. Lohr and R. J. Wyatt 1985. Did schizophrenia exist before the eighteenth century? *Comprehensive Psychiatry* 26: 493–503.
Kaskan, P. M. and B. L. Finley 2001. Encephalization and its developmental structure: how many ways can a brain get big. In D. Falk and K. R. Gibson (eds), *Evolutionary anatomy of the primate cerebral cortex*, pp. 14–29. Cambridge University Press, New York, NY.
Keller, M. C. and G. Miller 2006. Resolving the paradox of common, harmful, heritable mental disorders: which evolutionary genetic models work best? *Behavioral and Brain Sciences* 29: 385–452.
Kennedy, D. P., K. Semendeferi and E. Courchesne 2007. No reduction of spindle neuron number in the frontoinsular cortex in autism. *Brain and Cognition* 64(2): 124–129.
Kitano, T., Y. H. Liu, S. Ueda and N. Saitou 2004. Human-specific amino acid changes found in 103 protein-coding genes. *Molecular Biology and Evolution* 21: 936–944.
Kohl, P. L. and C. Fawcett 1995. Archaeology in the service of the state: theoretical considerations. In P. L. Kohl and C. Fawcett (eds), *Nationalism, politics, and the practice of archaeology*, pp. 3–18. Cambridge University Press, Cambridge.
Kolb, B. and I. Q. Whishaw 2008. *Fundamentals of human neuropsychology*, 6th edn. Worth Publishing Co., New York, NY.
Kuroki, N., M. E. Shenton, D. F. Salisbury, Y. Hirayasu, T. Onitsuka, H. Ersner-Hershfield, D. Yurgelun-Todd, R. Kikinis, F. A. Jolesz and R. W. McCarley 2006. Middle and inferior temporal gyrus gray matter volume abnormalities in first-episode schizophrenia: an MRI study. *American Journal of Psychiatry* 163(12): 2103–2110.
Leiner, H. C., A. L. Leiner and R. S. Dow 1995. Human brain map. In J. D. Schmahmann (ed), *The cerebellum and cognition*, p. 244. Academic, San Diego, CA.
Li, D., D. A. Collier and L. He 2006. Meta-analysis shows strong positive association of the neuregulin 1 (NRG1) gene with schizophrenia. *Human Molecular Genetics* 15: 1995–2002.
Lu, J., T. Tang, H. Tang, J. Huang, S. Shi and C.-I. Wu 2006. The accumulation of deleterious mutations in rice genomes: a hypothesis on the cost of domestication. *Trends in Genetics* 22: 126–131.
Lumsden, C. E. 1970. The neuropathology of multiple sclerosis. In P. J. Vinken and G. W. Bruyn (eds), *Handook of clinical neurology*, Vol. 9, pp. 217–309. North-Holland, Amsterdam.
Martin, R. 1996. Scaling of the mammalian brain: the maternal energy hypothesis. *News in the Physiological Sciences* 11: 149–156.
Mathalon, D. H. and J. M. Ford 2008. Divergent approaches converge on frontal lobe dysfunction in schizophrenia. *American Journal of Psychiatry* 154(8): 944–948.
Mayo Clinic Staff 2010. Frontotemporal dementia. Internet address: http://www.mayoclinic.com/health/frontotemporal-dementia/DS00874/METHOD=print
Mega, M. S. and J. Cummings 1997. The cingulate and cingulate syndromes. In M. R. Trimble and J. L. Cummings (eds), *Contemporary behavioural neurology*, pp. 189–214. Butterworths, Oxford.
Mueller, R.-A. and E. Courchesne 1998. The cerebellum: so much more. *Science* 181/5390: 879–880.
Muglia, P., A. Petronis, E. Mundo, S. Lander, T. Cate and J. L. Kennedy 2002. Dopamine D4 receptor and tyrosine hydroxylase genes in bipolar disorder: evidence for a role of DRD4. *Molecular Psychiatry* 7: 860–866.
Nimchimsky, E. A., E. Gilissen, J. M. Allman, D. P. Perl, J. M. Erwin and P. Hof 1999. A neuronal morphologic type unique to humans and great apes. *Proceedings of the National Academy of Sciences of the United States of America* 96: 5268–5273.
Nimchinsky, E. A., B. A. Vogt, J. H. Morrison and P. R. Hof 1995. Spindle neurons of the human anterior cingulate cortex. *Journal of Comparative Neurology* 355: 27–37.
Ochse, R. 1990. *Before the gates of excellence: the determinants of creative genius*. Cambridge University Press, Cambridge.
Olson, M. V. and A. Varki 2003. Sequencing the chimpanzee genome: insights into human evolution and disease. *Nature Reviews Genetics* 4: 20–28.

Polanyi, M. 1958. *Personal knowledge*. Routledge, London.

Preisig, M., F. Ferrero and A. Malafosse 2005. Monoamine oxidase a and tryptophan hydroxylase gene polymorphisms: are they associated in bipolar disorder? *American Journal of Pharmacogenetics* 5: 45–52.

Preuss, T. M. 2000. What's human about the human brain. In M. S. Gazzaniga (ed), *The New Cognitive Neurosciences*, pp. 1219–1234. MIT Press, Cambridge, MA.

Preuss, T. M. 2001. The discovery of cerebral diversity: an unwelcome scientific revolution. In D. Falk and K. R. Gibson (eds), *Evolutionary anatomy of the primate cerebral cortex*, pp. 138–164. Cambridge University Press, Cambridge.

Preuss, T. M. and G. Q. Coleman 2002. Human-specific organization of primary visual cortex: alternating compartments of dense Cat-301 and calbindin immunoreacitivity in layer 4A. *Cerebral Cortex* 12: 671–691.

Preuss, T. M. and J. H. Kaas 1999. Human brain evolution. In F. E. Bloom, S. C. Landes, J. L. Robert, L. R. Squire and M. J. Zigmond (eds), *Fundamental neuroscience*, pp. 1283–1311. Academic, San Diego, CA.

Rajkowska, G. 2009. Reductions in neuronal and glial density characterize the dorsolateral prefrontal cortex in bipolar disorder. *Biological Psychiatry* 49(9): 741–752.

Rakic, P. and D. R. Kornack 2001. Neocortical expansion and elaboration during primate evolution: a view from neuroembryology. In D. Falk and K. R. Gibson (eds), *Evolutionary anatomy of the primate cerebral cortex*, pp. 30–56. Cambridge University Press, New York, NY.

Randall, P. L. 1998. Schizophrenia as a consequence of brain evolution. *Schizophrenia Research* 30: 143–148.

Rathelot, R.-A. and P. L. Strick 2009. Subdivisions of primary motor cortex based on corticomotoneuronal cells. *Proceedings of the National Academy of Sciences of the United States of America* 106(3): 918–923.

Revilla, F. J. and J. Grutzendle 2008. Huntington disease. E-Medicine neurology specialities: movement and neurodegenerative disease. Internet address: http://emedicine.medscape.com/article/1150165-print

Rubinsztein, D. C., W. Amos, J. Leggo, S. Goodburn, R. S. Ramesar, J. Old, R. Dontrop, R. McMahon, D. E. Barton and, M. A. Ferguson-Smith 1994. Mutational bias provides a model for the evolution of Huntington's disease and predicts a general increase in disease prevalence. *Nature Genetics* 7(7): 525–530.

Sailor, M., B. Fischl, D. Salat, C. Tempelmann, E. Busa, N. Bodammer et al. 2003. Focal cortical thinning of the cerebral cortex in multiple sclerosis. *Brain* 126: 1734–1744.

Saito, T., F. Guan, D. F. Papolos, S. Lau, M. Klein, C. S. Fann and H. M. Lachman 2001. Mutation analysis of SYNJ1: a possible candidate gene for chromosome 21q22-linked bipolar disorder. *Molecular Psychiatry* 6: 387–395.

Sanjuan, J., A. Tolosa, J. C. Gonzalez, E. J. Aguilar, J. Perez-Tur, C. Najera, M. D. Molto and R. Frutos 2006. Association between FOXP2 polymorphisms and schizophrenia with auditory hallucinations. *Psychiatric Genetics* 16: 67–72.

Sapir, E. 1929. The status of linguistics as science. *Language* 5: 207–214.

Schildkraut, J. J. 1965. The catecholamine hypothesis of affective disorders: a review of supporting evidence. *American Journal of Psychiatry* 122: 509–528.

Schwartz, J. M. and S. Begley 2002. *The mind and the brain: neuroplasticity and the power of mental force*. Harper Collins, New York, NY.

Searle, J. R. 1995. *The construction of social reality*. Allen Lane, London.

Seeley, W. W., J. M. Allman, D. A. Carlin, R. K. Crawford, M. N. Macedo et al. 2007. Divergent social functioning in behavioral variant frontotemporal dementia and Alzheimer disease: reciprocal networks and neuronal evolution. *Alzheimer Disease and Associated Disorders* 21(94): S50–S57.

Seeley, W. W., D. A. Carlin and J. M. Allman 2006. Early frontotemporal dementia targets neurons unique to apes and humans. *Annals of Neurology* 60(6): 660–667.

Semendeferi, K. 1994. Evolution of the hominoid prefrontal cortex: a quantitative and image analysis of areas 13 and 10. PhD dissertation, University of Iowa.

Semendeferi, K. 1997. The evolution of the frontal lobes: a volumetric analysis based on three-dimensional reconstruction's of magnetic resonance scans of human and ape brains. *Journal of Human Evolution* 32: 375–388.

Semendeferi, K. 2001. Prefrontal cortex in humans and apes: a comparative study of area 10. *American Journal of Physical Anthropology* 114: 224–251.

Sherwood, C. C., J. K. Rilling, R. L. Holloway and P. R. Hof 2009. Evolution of the brain in humans: specializations in a comparative perspective. In M. D. Binder, N. Hirokawa and U. Windhorst (eds), *Encyclopedia of neuroscience*, pp. 1334–1338. Springer, Berlin-Heidelberg.

Silberman, N. A. 1995. Promised lands and chosen peoples: the politics and poetics of archaeological narrative. In P. L. Kohl and C. Fawcett (eds), *Nationalism, politics, and the practice of archaeology*, pp. 249–262. Cambridge University Press, Cambridge.

Smith, D. A. 2002. Imaging the progression of Alzheimer pathology through the brain. *Proceedings of the National Academy of Sciences of the United States of America* 99(7): 4135–4137.

Spinks, R., H. K. Sandhu, N. C. Andreasen and R. A. Philibert 2004. Association of the HOPA12 bp allele with a large X-chromosome haplotype and positive symptom schizophrenia. *American Journal of Medical Genetics, Part B: Neuropsychiatric Genetics* 127: 20–27.

Sridharan, D., D. J. Levitin and V. Menon 2008. A critical role for the right fronto-insular cortex in switching between central-executive and default-mode networks. *Proceedings of the National Academy of Sciences of the United States of America* 105(34): 12569–12574.

Stepniewska, I., T. M. Preuss and J. H. Kaas 1993. Architectonics, somatotopic organization and ipsilateral cortical connections of the primary motor area (M1) of owl monkeys. *Journal of Comparative Neurology* 330: 238–271.

Stopkova, P., J. Vevera, I. Paclt, I. Zukov and H. M. Lachman 2004. Analysis of SYNJ1, a candidate gene for 21q22 linked bipolar disorder: a replication study. *Psychiatry Research* 127: 157–161.

Tamminga, C. A., G. K. Thaker, R. Buchanan, B. Kirkpatrick et al. 1992. Limbic system abnormalities identified in schizophrenia using PET with fluorodeoxyglucose. *Archives of General Psychiatry* 49: 522–530.

Trigger, B. G. 1984. Alternative archaeologies: nationalist, colonialist, imperialist. *Man* 19: 355–370.

Trigger, B. G. 1985. The past as power: anthropology and the North American Indian. In I. McBryde (ed), *Who owns the past?*, pp. 11–40. Oxford University Press, Oxford.

Trigger, B. G. 1989. *A history of archaeological thought*. Cambridge University Press, Cambridge.

Ullian, E. M., B. T. Harris, A. Wu, J. R. Chan and B. A. Barres 2004. Schwann cells and astrocytes induce synapse formation by spinal motor neurons in culture. *Molecular Cell Neuroscience* 25(2): 241–251.

Ullian, E. M., S. K. Sapperstein, K. S. Christopherson and B. Barres 2001. Control of synapse number by glia. *Science* 291: 569–570.

Vawter, M. P., W. J. Freed and J. E. Kleinman 2000. Neuropathology of bipolar disorder. *Society of Biological Psychiatry* 48: 486–504.

Voight, B. F., S. Kudaravalli, X. Wen and J. K. Pritchard 2006. A map of recent positive selection in the human genome. *PLoS Biology* 4(3): e72.

Volle, F. O. and P. A. Heron-Helvenston 1979. *Epilepsy and you*. Charles C. Thomas Publishing Company, Springfield.

Walker, L. C. and L. C. Cork 1999. The neurobiology of aging in nonhuman primates. In R. D. Terry, R. Katzman, K. L. Bick and S. S. Sisodia (eds), *Alzheimer's disease*, 2nd edn., pp. 233–243. Lippincott Williams and Wilkins, Philadelphia, PA.

Watson, K. K., T. K. Kones and J. M. Allman 2006. Dendritic architecture of the von Economo neurons. *Neuroscience* 141: 1107–1112.

Wegner, C. and P. M. Mathews 2003. A new view of the cortex, new insights into multiple sclerosis. *Brain* 126: 1810–1821.

Wittgenstein, L. 1961. *Tractatus logico-philosophicus* (transl. by D. F. Pears and B. F. McGuinness). Routledge and Kegan Paul, London.

Wynn, T. 1979. The intelligence of later Acheulian hominids. *Man* 14: 371–492.

Wynn, T. 1989. *The evolution of spatial competence*. University of Illinois Press, Urbana, IL.

Wynn, T. 2002. Archaeology and cognitive evolution. *Behavioral and Brain Sciences* 25(3): 389–438.

Xu, M., D. St Clair and L. He 2006. Meta-analysis of association between ApoE epsilon4 allele and schizophrenia. *Schizophrenia Research* 84: 228–235.

Yoshikawa, T., M. Kikuchi, K. Saito, A. Watanabe, K. Yamada, H. Shibuya, M. Nankai, A. Kurumaji, E. Hattori, H. Ishiguro, H. Shimizu, Y. Okubo, M. Toru and S. D. Detera-Wadleigh 2001. Evidence for association of the myo-inositol monophosphatase 2 (IMPA2) gene with schizophrenia in Japanese samples. *Molecular Psychiatry* 6: 202–210.

Zhang, J., D. M. Webb and O. Podlaha 2002. Accelerated protein evolution and origins of human-specific features: FOXP2 as an example. *Genetics* 162: 1825–1835.

Index

R.G. Bednarik, *The Human Condition*, Developments in Primatology: Progress and Prospects, DOI 10.1007/978-1-4419-9353-3, © Springer Science+Business Media, LLC 2011